T0331553

Hyperbolicity
in Delay Equations

Series in Applied and Computational Mathematics

ISSN: 2010-2739

Series Editors: Philippe G. LeFloch *(Sorbonne University, France)*
Jian-Guo Liu *(Duke University, USA)*

Series in Applied and Computational Mathematics – Vol. 4

Hyperbolicity in Delay Equations

Luis Barreira
Claudia Valls

Universidade de Lisboa, Portugal

World Scientific

NEW JERSEY • LONDON • SINGAPORE • BEIJING • SHANGHAI • HONG KONG • TAIPEI • CHENNAI • TOKYO

Published by

World Scientific Publishing Co. Pte. Ltd.

5 Toh Tuck Link, Singapore 596224

USA office: 27 Warren Street, Suite 401-402, Hackensack, NJ 07601

UK office: 57 Shelton Street, Covent Garden, London WC2H 9HE

Library of Congress Cataloging-in-Publication Data
Names: Barreira, Luis, 1968– author. | Valls, Claudia, 1973– author.
Title: Hyperbolicity in delay equations / Luis Barreira, Claudia Valls.
Description: Hackensack, New Jersey : World Scientific, [2021] |
 Series: Series in applied and computational mathematics, 2010-2739 ; vol. 4 |
 Includes bibliographical references and index.
Identifiers: LCCN 2020051743 | ISBN 9789811230240 (hardcover) |
 ISBN 9789811230257 (ebook for institutions) | ISBN 9789811230264 (ebook for individuals)
Subjects: LCSH: Delay differential equations. | Differentiable dynamical systems. | Stability.
Classification: LCC QA371 .B297 2021 | DDC 515/.35--dc23
LC record available at https://lccn.loc.gov/2020051743

British Library Cataloguing-in-Publication Data
A catalogue record for this book is available from the British Library.

For any available supplementary material, please visit
https://www.worldscientific.com/worldscibooks/10.1142/12098#t=suppl

Printed in Singapore

Preface

The main objective of this book is to give a comprehensive introduction to the study of *hyperbolicity* in delay equations, both linear and nonlinear. This includes a self-contained discussion of the foundations, main results and main techniques, with emphasis on those parts of the theory that are of sufficiently general nature and which thus apply to a large class of delay equations. The main theme is always hyperbolicity and on purpose we avoided discussing topics that are not directly related to it.

The book is dedicated to researchers as well as graduate students specializing in differential equations and dynamical systems who wish to have a sufficiently comprehensive view of the hyperbolicity theory of delay equations. It can also be used as a basis for graduate courses on the stability and hyperbolicity of delay equations.

The material is divided into four parts: prelude, linear stability, nonlinear stability, and further topics. We emphasize that with the exception of a few results from other areas, everything in the book is proved.

Part I is an introduction to the foundations of the theory of delay equations and to the hyperbolicity theory. It includes all the necessary material from both theories that is used in the book, thus making the text self-contained. The topics include the existence, uniqueness and continuation of solutions under the Carathéodory conditions, the variation of constants formula, and finally the notion of an exponential dichotomy.

Part II is dedicated to the stability of a linear delay equation under sufficiently small linear perturbations. The topics include the robustness of an exponential dichotomy on the line and the notion of admissibility together with its characterization of the existence of an exponential dichotomy and its applications to the robustness of hyperbolicity, both on the line and on the half-line.

Part III is dedicated to the construction of stable and unstable invariant manifolds for any sufficiently small perturbation of a linear delay equation with an exponential dichotomy. We first consider Lipschitz perturbations and we construct Lipschitz invariant manifolds. We also establish a partial version of the Grobman–Hartman theorem by linearizing the unstable component of the dynamics. We then consider the construction of smooth invariant manifolds for any sufficiently small smooth perturbation of an exponential dichotomy.

Finally, in Part IV we consider a few additional topics that also play an important role in the theory. In particular, we construct smooth center invariant manifolds for any sufficiently small smooth perturbation of an exponential trichotomy. We also consider the spectrum of a linear delay equation and we give a description of all its possible forms. Finally, we consider nonlinear perturbations of a linear delay equation and we give assumptions under which the Lyapunov exponents of the nonlinear dynamics are close to those of the linear dynamics.

Luis Barreira and Claudia Valls

Contents

Linear Stability 71

Nonlinear Stability 127

PART I
Prelude

Chapter 1

Introduction

We start this introductory chapter with a brief nontechnical introduction to the area of delay equations, without an effort of being comprehensive. We also include a description of the topics considered in the book as well as pointers for further developments. The main theme is always hyperbolicity, in the context of delay equations, both linear and nonlinear. The specific topics include the persistence of stability and conditional stability under sufficiently small linear and nonlinear perturbations, the robustness property, the admissibility property, the construction of invariant manifolds, and the study of the spectral properties of linear and nonlinear delay equations.

1.1 Delay Equations

Brief History of the Area

Delay equations model phenomena for which the present state depends on some or all of the past states. They have been studied for quite some time, but it was not until the 1950's and particularly the 1960's that the theory gained the explosive momentum and finally reached the modern approach that still characterizes it today. In particular, some isolated studies of specific delay equations, going back to the late 18th century, are due to Laplace, Poisson, Condorcet, Lacroix, Boole, etc. According to Bellman and Cooke [15], the first authors to consider general classes of delay equations were Polossuchin [114] and Schmidt [123] (in fact the second work is a generalization of the first one).

A passionate exposition of the important role that the study of delay equations could have or should have in the study of life sciences was given by Picard in 1908 [109]. In particular, he noted that while mechanical phenomena carry a more or less explicit postulate of nonheredity, according

3

to which one expects that the movement of an object can be determined completely after knowing the positions and velocities at the present instant, in life sciences many phenomena keep traces of the past states. He also remarked that although it was not known at the time whether mathematics could at some point be used to study such natural phenomena, one could "dream" of being able to consider functional equations containing integrals over past times. These integrals were precisely the hereditary components.

This "invitation" made by Picard was followed systematically by Volterra who were among the first to take sound steps towards initiating a mathematical theory of delay equations, despite strong voices against it from the conceptual point of view. Since it is not our aim to give a detailed account of the history of the theory, we shall content ourselves with recalling Volterra's own opinion on his contributions to integro-differential equations [138]:

> "A questo punto viene fatto di domandarsi se i fenomeni naturali si esplichino realmente così, o se non sia invece da supporre che esista effettivamente una eredità dei fatti passati [...] Credo di aver dimostrato che le teorie ereditarie, almeno quelle di cui ho parlato, possano essere studiate mediante equazioni integro-differenziali; si può così sviluppare lo studio teorico dell'eredità senza fare alcuna ipotesi speciale sulle funzioni che la caratterizzano, ossia sui coefficienti di eredità."

that translates to:

> At this point we wonder whether natural phenomena can really be explained like this, or whether we should not assume that there actually exists a dependence of past events [...] I think that I have shown that the hereditary theory, at least the part that I have considered, can be studied using integro-differential equations; in this way one can develop the theoretical study of heredity without making any special assumptions about the functions that characterize it, that is, about the coefficients of heredity.

During the 1950's there was an explosion of activity, culminating with the books of Myshkis [95] in 1951 (where one can see the classification of linear delay equations into retarded, neutral and advanced), Krasovskiĭ [70] in 1959, Bellman and Cooke [15] in 1963 (building on the former book of Bellman and Danskin [16]) and Halanay [47] in 1966. These works give a panorama of the huge development of the theory up to the mid 1960's. However, it still lacked the modern approach that characterizes it today.

For example, we can find the variation of constants formula in [15] (with the usual discontinuous initial condition), but the language and point of view are far from those that started being used more or less at the same time and which helped substantially in the development of the theory.

Indeed, in the late 1950's and early 1960's, Hale started working on delay equations following the approach proposed by Krasovskiĭ in 1956 [69], which included considering the solutions taking as phase space the continuous functions. In his seminal work [48] Hale wrote:

> "Until recently, most of the results concerning differential-difference equations have been obtained by treating the dependent variable as a point in Euclidean space and employing arguments which are standard in the theory of ordinary differential equations. To the author's knowledge, Krasovskii [70] was the first to exploit the idea that the proper setting for these problems is in a function space. In doing so, the arguments used for ordinary differential equations become more natural for differential-difference equations. The present paper is an attempt to obtain some analogies between linear differential-difference equations with constant coefficients and ordinary linear differential equations with constant coefficients."

The same paper contains a description of the eigenspaces associated with an autonomous linear delay equation $v' = Lv_t$, where $L \colon C \to \mathbb{R}^n$ is a bounded linear operator acting on the set C of all continuous functions $\varphi \colon [-r, 0] \to \mathbb{R}^n$, as well as a discussion of the behavior on each eigenspace and how many perturbation results in the theory of ordinary differential equations extend easily to delay equations.

Since the works of Volterra on predator-prey models and viscoelasticity, together with the great interest in delayed feedback mechanisms in engineering problems and control theory, the theory of delay equations experienced a rapid and exciting development. We refer the reader to Hale's book [50] for the first comprehensive introduction to the area as a consolidated mathematical theory (building on his former work [49]) and to the books [34, 52, 54, 140] for further developments.

Basic Notions and Results

As described in the preface, our book is essentially dedicated to the study of hyperbolicity, in the context of delay equations. We first recall some basic notions and results, including the usual theorems of existence, uniqueness and continuation of solutions, always under the Carathéodory conditions.

Let $|\cdot|$ be a norm on $\mathbb{R}^n$. Given $r \geq 0$, we denote by C the Banach space of all continuous functions $\varphi \colon [-r, 0] \to \mathbb{R}^n$ with the norm

$$\|\varphi\| = \sup\{|\varphi(\theta)| : -r \leq \theta \leq 0\}.$$

Now let $f \colon \Omega \to \mathbb{R}^n$ be a function satisfying the usual Carathéodory conditions in the theory of ordinary differential equations (see Definition 2.1) on an open set $\Omega \subset \mathbb{R} \times C$. We consider the *delay equation*

$$v' = f(t, v_t), \tag{1.1}$$

where v' denotes the right-hand derivative (when it exists) and $v_t(\theta) = v(t + \theta)$ for $\theta \in [-r, 0]$, assuming that v is defined at least on the interval $[t - r, t]$. A continuous function $v \colon [s - r, a) \to \mathbb{R}^n$ with $a \leq +\infty$ is called a *solution* of Eq. (1.1) if $(t, v_t) \in \Omega$ and

$$v(t) = v(s) + \int_s^t f(\tau, v_\tau)\, d\tau \quad \text{for } t \in [s, a).$$

One can also consider solutions with domain $(-\infty, a)$ (see Definition 2.3).

Given an initial condition, that is, a pair $(s, \varphi) \in \Omega$, we consider the *initial value problem*

$$\begin{cases} v' = f(t, v_t), \\ v_s = \varphi. \end{cases} \tag{1.2}$$

Problem (1.2) consists of finding a solution $v \colon [s - r, a) \to \mathbb{R}^n$ of Eq. (1.1), for some $a > s$, such that $v_s = \varphi$.

We start by formulating a result that considers the existence of solutions, their uniqueness under a Lipschitz property of f and their continuation (see Sec. 2.1 for details).

Theorem 1.1. *Given a function $f \colon \Omega \to \mathbb{R}^n$ satisfying the Carathéodory conditions on an open set $\Omega \subset \mathbb{R} \times C$ and a pair $(s, \varphi) \in \Omega$:*

(1) there exists a solution of the initial value problem (1.2);
(2) if f is locally Lipschitz in the second variable, then the initial value problem (1.2) has a unique solution on some interval $[s - r, a)$;
(3) given a noncontinuable solution v of Eq. (1.1) on $[s - r, a)$, for each compact set $W \subset \Omega$ there exists $t_W \in [s - r, a)$ such that $(t, v_t) \notin W$ for $t \in [t_W, a)$.

One can also consider delay equations whose solutions have values in a Banach space X, that is, functions $f \colon \Omega \to X$ satisfying the Carathéodory conditions on an open set $\Omega \subset \mathbb{R} \times C$. In this case, one can show that if

$v\colon [s-r,a) \to X$ is a solution of Eq. (1.1), then v is absolutely continuous on $[s,a)$ and $v' = f(t,v_t)$ for almost every $t \in [s,a)$. The converse statement holds for any reflexive Banach space X (and so in particular for any finite-dimensional space), but not in general (see [68]).

In the text we avoided discussing delay equations with infinite delay. We refer the reader to the book [61] for a detailed treatment of various topics in that setting. See also [127] for the study of delay equations with infinite delay whose solutions have values in an arbitrary Banach space.

Linear Equations

In Secs. 2.2 and 2.3 we consider at length the particular class of nonautonomous linear delay equations. We emphasize that all nonlinear equations considered in the book are (seen as) perturbations of linear equations and so it is crucial that linear equations are discussed in detail, including their hyperbolicity. In particular, we introduce their evolution families on the space C and we show how these can be extended to certain discontinuous functions.

Let $L(t)\colon C \to \mathbb{R}^n$, for $t \in \mathbb{R}$, be bounded linear operators and assume that the function $(t,v) \mapsto L(t)v$ satisfies the Carathéodory conditions on $\Omega = \mathbb{R} \times C$ and that

$$\int_s^t \|L(\tau)\|\,d\tau < +\infty \quad \text{for } t > s. \tag{1.3}$$

We consider the *linear delay equation*

$$v' = L(t)v_t. \tag{1.4}$$

Under the former assumptions, Eq. (1.4) has unique forward global solutions (see Theorem 2.4). Hence, it determines an evolution family $T(t,s)$ of bounded linear operators on C defined by

$$T(t,s)\varphi = v_t \tag{1.5}$$

for $t \geq s$ and $\varphi \in C$, where v is the unique solution of the equation on $[s-r,+\infty)$ with $v_s = \varphi$. This means that $T(t,t) = \text{Id}$ and

$$T(t,s) = T(t,\tau)T(\tau,s)$$

for all $t \geq \tau \geq s$. An important property when $r > 0$ is that each $T(t,s)$ is compact whenever $t \geq s + r$ (see Theorem 2.5).

In fact, one can extend the linear operators $T(t,s)$ to a certain space of discontinuous functions. Let C_0 be the set of all functions $\varphi\colon [-r,0] \to \mathbb{R}^n$ that are continuous on $[-r,0)$ for which the limit

$$\varphi(0^-) = \lim_{\theta \to 0^-} \varphi(\theta)$$

exists. This is a Banach space when equipped with the supremum norm. Now we write each operator $L(t)\colon C \to \mathbb{R}^n$ as a Riemann–Stieltjes integral

$$L(t)\varphi = \int_{-r}^{0} d\eta(t,\theta)\varphi(\theta) = \int_{-r}^{0} d_\theta\eta(t,\theta)\varphi(\theta) \qquad (1.6)$$

for some measurable map

$$\eta\colon \mathbb{R} \times [-r,0] \to M_n,$$

where M_n is the set of all $n \times n$ matrices, such that $\theta \mapsto \eta(t,\theta)$ has bounded variation and is left-continuous for each $t \in \mathbb{R}$ (the symbol d_θ means that the integral is taken with respect to θ). Since each function $\varphi \in C_0$ is right-continuous, it is Riemann–Stieltjes integrable with respect to $\eta(t,\cdot)$ for each $t \in \mathbb{R}$. Hence, one can extend each linear operator $L(t)$ to C_0 using the integral in (1.6). We continue to denote the extension by $L(t)$. One can verify that

$$\|L(t)|C\| = \|L(t)|C_0\|.$$

Given $t, s \in \mathbb{R}$ with $t \geq s$, we define a linear operator $T_0(t,s)$ on C_0 by

$$T_0(t,s)\varphi = v_t$$

for $\varphi \in C_0$, where v is the unique solution of Eq. (1.4) on $[s - r, +\infty)$ with $v_s = \varphi$ (see Theorem 2.6). We refer to Sec. 2.3 for details and proofs.

One can also consider linear delay equations whose solutions have values in an arbitrary Banach space X, but we shall not pursue the corresponding theory in the book. In particular, a linear operator $L(t)\colon C \to X$ with values in a Banach space X can also be represented as in (1.6) for some measurable map

$$\eta\colon \mathbb{R} \times [-r,0] \to \mathcal{L}(X)$$

such that $\theta \mapsto \eta(t,\theta)$ has bounded variation and is left-continuous for each $t \in \mathbb{R}$ (see [133]), where $\mathcal{L}(X)$ denotes the set of all bounded linear operators from X into itself.

Variation of Constants Formula

Many of the results proven in the book use the variation of constants formula for the perturbations of a nonautonomous linear delay equation. Namely, we consider the perturbed equation

$$v' = L(t)v_t + g(t), \qquad (1.7)$$

with the operator $L(t)$ as above and where $g\colon \mathbb{R} \to \mathbb{R}^n$ is a locally integrable function. Given $(s, \varphi) \in \Omega$, there exists a unique forward global solution $v\colon [s - r, +\infty) \to \mathbb{R}^n$ of Eq. (1.7) satisfying the initial condition $v_s = \varphi$ (see Proposition 2.5).

Now we define a linear operator $X_0\colon \mathbb{R}^n \to C_0$ by

$$(X_0 p)(\theta) = \begin{cases} 0 & \text{if } -r \le \theta < 0, \\ p & \text{if } \theta = 0 \end{cases}$$

for each $p \in \mathbb{R}^n$. The *variation of constants formula* (see Theorem 2.7) shows that a function $v\colon [s-r, +\infty) \to \mathbb{R}^n$ is the unique solution of Eq. (1.7) with $v_s = \varphi$ if and only if

$$v_t = T(t, s)\varphi + \int_s^t T_0(t, \tau) X_0 g(\tau)\, d\tau \quad \text{for } t \ge s.$$

The latter identity is to be understood in the sense that $v_s = \varphi$ and

$$v(t + \theta) = (T(t, s)\varphi)(\theta) + \int_s^{t+\theta} \big(T_0(t, \tau) X_0 g(\tau)\big)(\theta)\, d\tau$$

for all $t \ge s$ and $\theta \in [-r, 0]$ with $t + \theta \ge s$. In Sec. 2.4 we give a proof of the variation of constants formula using only the results established in the book.

Finally, in Sec. 2.5 we consider the simpler class of autonomous linear delay equations and their nonautonomous perturbations. An autonomous linear delay equation $v' = Lv_t$, where $L\colon C \to \mathbb{R}^n$ is a bounded linear operator, generates a semigroup $S(t)$ such that

$$T(t, s) = S(t - s) \quad \text{for } t \ge s. \tag{1.8}$$

This causes that the corresponding variation of constants formula is somewhat simpler: namely, a function $v\colon [s-r, +\infty) \to \mathbb{R}^n$ is the unique solution of the equation

$$v' = Lv_t + g(t)$$

with $v_s = \varphi$ if and only if

$$v_t = S(t - s)\varphi + \int_s^t S_0(t - \tau) X_0 g(\tau)\, d\tau \quad \text{for } t \ge s,$$

where $S_0(t - \tau) = T_0(t, \tau)$.

1.2 Hyperbolicity and Robustness

The Notion of Hyperbolicity

The notion of an exponential dichotomy, essentially introduced by Perron in [105], plays a central role in a large part of the theory of differential equations and dynamical systems. In particular, it is central in the stability theory, such as in connection with the construction of topological conjugacies, invariant manifolds and normal forms. For example, the existence of an exponential dichotomy for a linear delay equation $v' = L(t)v_t$ leads to the existence of stable and unstable invariant manifolds for the nonlinear delay equation

$$v' = L(t)v_t + g(t, v_t)$$

under mild additional assumptions on the nonlinear perturbation g (see Chaps. 7 and 8). Moreover, the local instability of the trajectories caused by the existence of an exponential dichotomy influences the global behavior of the system. This instability is one of the main mechanisms responsible for the occurrence of chaotic behavior. For details and references on hyperbolicity and its applications we refer the reader to [32, 51, 58, 87, 125] and specifically to [34, 50, 54] for delay equations.

Now we consider the specific case of delay equations. Namely, let $L(t) \colon C \to \mathbb{R}^n$, for $t \in \mathbb{R}$, be bounded linear operators and consider the linear delay Eq. (1.4). We assume that $t \mapsto L(t)\varphi$ is measurable for each $\varphi \in C$ and that $t \mapsto \|L(t)\|$ is locally integrable, with

$$\sup_{t \in \mathbb{R}} \int_t^{t+1} \|L(\tau)\| \, d\tau < +\infty. \tag{1.9}$$

Let $T(t, s)$ be the evolution family defined by (1.5). We say that Eq. (1.4) has an *exponential dichotomy on an interval* $I \subset \mathbb{R}$ if:

(1) there exist projections $P(t) \colon C \to C$, for $t \in I$, such that

$$P(t)T(t, s) = T(t, s)P(s) \quad \text{for } t \geq s;$$

(2) the linear operator

$$\overline{T}(t, s) = T(t, s)|\ker P(s) \colon \ker P(s) \to \ker P(t)$$

is invertible for $t \geq s$;

(3) there exist $\lambda, D > 0$ such that for every $t, s \in I$ with $t \geq s$ we have

$$\|T(t, s)P(s)\| \leq De^{-\lambda(t-s)}, \quad \|\overline{T}(s, t)Q(t)\| \leq De^{-\lambda(t-s)},$$

where $\overline{T}(s, t) = \overline{T}(t, s)^{-1}$ and $Q(t) = \mathrm{Id} - P(t)$.

Then the spaces

$$E(t) = P(t)C \quad \text{and} \quad F(t) = Q(t)C$$

are called, respectively, *stable* and *unstable spaces* at time t. An important property of an exponential dichotomy on an interval I containing $\mathbb{R}_0^+$ is that the unstable spaces are finite-dimensional when $r > 0$ (see Theorem 3.2). This is a simple consequence of the compactness of the linear operators $T(t,s)$ for $t \geq s + r$.

In Sec. 3.2 we show how the existence of an exponential dichotomy can be characterized in terms of strict Lyapunov functions. This provides an alternative approach to verify the existence of exponential behavior. We first consider exponential contractions, in which case $P(t) = \text{Id}$ for all $t \in I$.

Moreover, in Sec. 3.5 we consider the particular case of an autonomous linear differential equation $v' = Lv_t$. Then the semigroup $S(t)$ in (1.8) is strongly continuous and we define its generator $A: D(A) \to C$ by

$$A\varphi = \lim_{t \searrow 0} \frac{S(t)\varphi - \varphi}{t}$$

in the domain $D(A) \subset C$ of all $\varphi \in C$ for which the limit exists in the norm topology of C. After proving some properties of A (first established by Hale in [48]), we show that if the spectrum $\sigma(A)$ does not intersect the imaginary axis, then the equation $v' = Lv_t$ has an exponential dichotomy on $\mathbb{R}$ (see Theorem 3.8). We note that we give a streamlined proof that avoids the adjoint equation.

Before proceeding, we note that a main aspect of the notion of an exponential dichotomy is the contraction and expansion along certain invariant subspaces. But there is a second aspect, often not stressed sufficiently: the angles between the stable and unstable spaces (or the norms of the corresponding projections, in the infinite-dimensional case) are uniformly bounded away from zero. These two aspects can be weaken to a considerable extent, leading to the much more general notion of *nonuniform hyperbolicity*, which has many applications, for example in the study of dynamical systems with nonzero Lyapunov exponents and particularly in smooth ergodic theory (see for example [3, 4]). Namely:

(1) the contraction and expansion along the stable and unstable spaces need not be uniform;
(2) the angles between the stable and unstable spaces may approach zero along orbits, although not exponentially.

In comparison with the notion of an exponential dichotomy, this weaker notion of hyperbolicity is a much weaker requirement. In particular, from the point of view of ergodic theory *almost all* linear variational equations with nonzero Lyapunov exponents have a nonuniform exponential behavior. More precisely, for an autonomous differential equation $x' = F(x)$ whose flow φ_t preserves a finite measure, for almost all x the linear variational equation

$$y' = A(t)y, \quad \text{with } A(t) = d_{\varphi_t(x)}F,$$

has a nonuniform exponential dichotomy when all its Lyapunov exponents are nonzero.

We refer the reader to [4] for a comprehensive introduction to the theory, which goes back to works of Oseledets [99] and particularly Pesin [106–108]. In [85] Mañé established related results on nonuniform hyperbolicity, Lyapunov exponents and stable invariant manifolds that apply to some classes of delay equations. The study of the notion of nonuniform hyperbolicity is already out of the scope of our book, particularly due to the need for much additional material from ergodic theory.

Decompositions and Upper Bounds

It turns out that if Eq. (1.4) has an exponential dichotomy on an interval I containing $\mathbb{R}_0^+$, then the extension $T_0(t, s)$ of $T(t, s)$ to the space C_0 behaves as if we also had an exponential dichotomy on I. Essentially this amounts to decompose the operator X_0 into stable and unstable components. More precisely, for each $t \in I$ we define linear operators $P_0(t), Q_0(t) \colon \mathbb{R}^n \to C_0$ by

$$P_0(t) = X_0 - Q_0(t)$$

and

$$Q_0(t) = \overline{T}(t, t + r)Q(t + r)T_0(t + r, t)X_0.$$

One can show that

$$P_0(t)p \in C_0 \setminus C \quad \text{and} \quad Q_0(t)p \in C$$

for all $p \in \mathbb{R}^n$. Assuming that condition (1.9) holds, if Eq. (2.12) has an exponential dichotomy on an interval I containing $\mathbb{R}_0^+$, then there exist constants $\lambda, N > 0$ such that

$$\|T_0(t, s)P_0(s)\| \le Ne^{-\lambda(t-s)}, \quad \|\overline{T}(s, t)Q_0(t)\| \le Ne^{-\lambda(t-s)}$$

for every $t, s \in I$ with $t \ge s$ (see Theorem 3.4).

An important consequence of this extension of an exponential dichotomy to the space C_0 is that the variation of constants formula also projects onto the stable and unstable spaces. Namely, consider the perturbed Eq. (1.7), with the same assumptions as before. Assuming that Eq. (1.4) has an exponential dichotomy on an interval I containing $\mathbb{R}_0^+$, for each $(s, \varphi) \in \Omega$ the unique solution $v \colon [s - r, +\infty) \to \mathbb{R}^n$ of Eq. (1.7) with $v_s = \varphi$ satisfies

$$P(t)v_t = T(t, s)P(s)\varphi + \int_s^t T_0(t, \tau)P_0(\tau)g(\tau)\,d\tau$$

and

$$Q(t)v_t = T(t, s)Q(s)\varphi + \int_s^t T(t, \tau)Q_0(\tau)g(\tau)\,d\tau$$

for all $t \geq s$ (see Theorem 3.6).

Robustness of Exponential Dichotomies

The relevance of the notion of hyperbolicity in a large part of the theory of differential equations and dynamical systems makes it crucial to understand whether the exponential behavior persists under sufficiently small perturbations. In this context we are also interested in the study of the robustness property of an exponential dichotomy for a nonautonomous linear delay equation. Having this property means that the existence of an exponential dichotomy for a given linear equation persists under sufficiently small linear perturbations. We emphasize that the study of robustness (particularly in the context of ordinary differential equations) has a long history. In particular, it was discussed by Massera and Schäffer [86] (building on [105]), Coppel [31] and in the case of Banach spaces by Dalec'kiĭ and Kreĭn [33].

Let $L(t), M(t) \colon C \to \mathbb{R}^n$, for $t \in \mathbb{R}$, be bounded linear operators. We assume that the functions $(t, v) \mapsto L(t)v$ and $(t, v) \mapsto M(t)v$ satisfy the Carathéodory conditions on $\Omega = \mathbb{R} \times C$ and that (1.3) holds. We also consider the perturbed equation

$$v' = (L(t) + M(t))v_t. \tag{1.10}$$

The following theorem is a corollary of a robustness result in Theorem 4.1 that was obtained in [10].

Theorem 1.2. *Assume that Eq. (1.4) has an exponential dichotomy on $\mathbb{R}$ and that there exists $\delta > 0$ such that $\|M(t)\| \leq \delta$ for all $t \in \mathbb{R}$. If δ is sufficiently small, then Eq. (1.10) has also an exponential dichotomy on $\mathbb{R}$.*

In Theorem 4.1 we consider the more general assumption

$$\sup_{t \in \mathbb{R}} \int_t^{t+1} \|M(\tau)\| \, d\tau \leq \delta, \tag{1.11}$$

which relates better to the integrability assumption in (1.9). In addition, we show that:

(1) the stable and unstable spaces $P(t)C$ and $Q(t)C$ for Eq. (1.4) are isomorphic, respectively, to the stable and unstable spaces for the perturbed Eq. (1.10);

(2) given $\varepsilon > 0$, if δ is sufficiently small and λ is the exponent in the exponential dichotomy for Eq. (1.4), then $\lambda - \varepsilon$ is an exponent in the exponential dichotomy for Eq. (1.10).

Our approach exhibits in a more or less explicit manner the stable and unstable spaces for the perturbed equation as images of linear operators obtained through appropriate fixed point problems. The construction of these linear operators essentially corresponds to the construction of bounded solutions into the future and into the past (recall that the stable and unstable spaces are formed by the initial conditions leading to bounded solutions, respectively, into the future and into the past). The rest of the argument consists of obtaining exponential bounds along the stable and unstable spaces of the perturbed equation as well as bounds for the norms of the projections onto these spaces. To the possible extent we follow the approach developed in [5] in the context of ordinary differential equations on Banach spaces (some of the arguments are inspired on work of Popescu in [115]). The latter was then extended to delay equations in [10], on which our presentation is inspired.

In the case of delay equations, Lizana [81] stated a robustness result without proof, under the more restrictive assumptions that the linear operators $L(t)$ and $M(t)$ are bounded and uniformly continuous in t and that $\sup_{t \in \mathbb{R}} \|M(t)\|$ is sufficiently small. We note that these assumptions are a special case of the assumptions in Theorem 4.1. More precisely, we consider equations satisfying the Carathéodory conditions and the boundedness of the perturbation in the supremum norm is replaced by condition (1.11).

There are various other methods to establish the robustness of an exponential dichotomy. We describe a few that are based on a characterization of the notion, after which the argument consists of showing that a perturbation also satisfies the properties in the characterization. In this respect, our approach differs substantially from any of these methods since we do not need any characterization of the notion of an exponential dichotomy. Instead, as noted above, we construct directly the stable and unstable spaces for the perturbed equation:

(1) *Admissibility.* The notion of admissibility also goes back to Perron and referred originally to the characterization of the hyperbolicity of a linear differential equation $x' = A(t)x$ in terms of the existence and uniqueness of bounded solutions of the equation

$$x' = A(t)x + f(t) \tag{1.12}$$

for any bounded continuous perturbation f. One can also consider other spaces where we take the perturbations and look for the solutions (such as for example L^p spaces). In addition, one can consider strong and weak solutions. For details and references we refer the reader to the books [2, 23]. We mention in particular the works of Chow and Leiva [26] and Pliss and Sell [113], where the authors study the robustness property in the context of skew-product semiflows over a compact base. They discretize the problem and then use admissibility results to complete the argument (also building on work of Henry [58]).

(2) *Lyapunov functions.* According to Coppel [31], the relation between Lyapunov functions and exponential dichotomies was first considered by Maĭzel' in [84]. Related approaches were described by Dalec'kiĭ and Kreĭn [33] and Massera and Schäffer [87]. The use of Lyapunov functions in the study of the stability of solutions of differential equations goes back to Lyapunov (see [82]). Among the first accounts of the theory are the books [17,46,74]. We refer to [93] for a detailed discussion of the relation between Lyapunov functions and exponential dichotomies. We note that in the context of ergodic theory there is also a related approach, going back to Wojtkowski [139]. We refer to [4] for details and further references.

(3) *Fredholm alternative.* To obtain a criterion for the existence of an exponential dichotomy one can also use a Fredholm alternative, sometimes in connection with admissibility. For a linear equation $x' = A(t)x$ this consists of giving a description of the existence of solutions of Eq. (1.12) in terms of those of the adjoint equation $y' = -A(t)^*y$. Related work is due to Palmer [101] for ordinary differential equations, Lin [79] for functional differential equations, Blázquez [18], Rodrigues and Silveira [120], Zeng [141] and Zhang [142] for parabolic evolution equations, and Chow and Leiva [26], Sacker and Sell [122] and Rodrigues and Ruas-Filho [119] for abstract evolution equations.

1.3   Admissibility and Hyperbolicity

As already noted in the former section, the notion of admissibility, essentially introduced by Perron in [105], referred originally (in the context of ordinary differential equations) to the existence and uniqueness of bounded solutions of the perturbations

$$x' = A(t)x + f(t) \tag{1.13}$$

of a linear equation

$$x' = A(t)x \tag{1.14}$$

for any bounded continuous function f. This property can be used to deduce the stability or the conditional stability of the linear equation under sufficiently small linear and nonlinear perturbations. Incidentally, Perron's paper not only contributed to prepare the ground for the notion of an exponential dichotomy, but also for the stable manifold theorem. In this spirit, we recall a particular result from [105] (or, more precisely, a simple consequence of one of the results). Let $A(t)$ be $n \times n$ matrices varying continuously with $t \in \mathbb{R}$.

Theorem 1.3. *If Eq. (1.13) has at least one bounded solution on $\mathbb{R}_0^+$ for each bounded continuous function f, then each bounded solution of Eq. (1.14) tends to zero when $t \to +\infty$.*

The assumption in Theorem 1.3 is called the *admissibility* of the pair of spaces in which we take the perturbations and look for the solutions. The result is probably the first place in the literature where one can see a relation between admissibility and stability.

We note that one can also consider the admissibility of other pairs of spaces, such as (L^p, L^q) for $p, q \in [1, \infty)$ with $p \geq q$. Then the (strong) admissibility property corresponds to require that for each $f \in L^q$ there exists a unique $v \in L^p$ that is absolutely continuous on each compact interval and satisfies identity (1.13) for Lebesgue-almost every t. In addition, one can consider a (weak) admissibility property, which is defined in terms of the existence of mild solutions for the corresponding integral equation.

There is an extensive literature on the relation between admissibility and stability, also in infinite-dimensional spaces. For a description of the most relevant early contributions in the area we refer to the books by Massera and Schäffer [87] (building on their former work [86]) and by Dalec'kiĭ and Kreĭn [33]. See [78] for some early results in infinite-dimensional spaces.

For details and references we refer the reader to the book [23] (see also [63, 91, 92, 96, 116, 134] and the references therein). See [26, 113] for related results for skew-product semiflows over a compact base.

In Chap. 5 we show that the notion of an exponential dichotomy for a nonautonomous linear delay equation can also be characterized in terms of an admissibility property. Namely, after introducing appropriate Banach spaces where we take the perturbations of the linear equation and where we look for the solutions, we show that the existence of an exponential dichotomy is equivalent to the admissibility of this pair of spaces. Moreover, we consider delay equations on the line and on the half-line.

We formulate briefly the results. We start with the case of the interval $I = \mathbb{R}$. Let $\mathcal{E}$ be the set of all bounded continuous functions $v\colon \mathbb{R} \to \mathbb{R}^n$ and let $\mathcal{F}$ be the set of all measurable functions $g\colon \mathbb{R} \to \mathbb{R}^n$ such that

$$\sup_{t\in\mathbb{R}} \int_t^{t+1} |g(\tau)|\, d\tau < +\infty,$$

identified if they are equal almost everywhere. The following result is a combination of Theorems 5.1 and 5.2 that were obtained in [12].

Theorem 1.4. *Equation* (1.4) *has an exponential dichotomy on* $\mathbb{R}$ *if and only if for each* $g \in \mathcal{F}$ *there exists a unique* $v \in \mathcal{E}$ *such that*

$$v_t = T(t,s)v_s + \int_s^t T_0(t,\tau)X_0 g(\tau)\, d\tau \qquad (1.15)$$

for $t, s \in \mathbb{R}$ *with* $t \geq s$.

Now we consider the case of the interval $I = \mathbb{R}_0^+$, which requires a different notion of admissibility. Let $\mathcal{E}^+$ be the set of all bounded continuous functions $v\colon [-r, +\infty) \to \mathbb{R}^n$ and given a closed subspace $B \subset C$, let $\mathcal{E}_B^+$ the set of all functions $v \in \mathcal{E}^+$ such that $v_0 \in B$. Finally, let $\mathcal{F}^+$ be the set of all measurable functions $g\colon [-r, +\infty) \to \mathbb{R}^n$ such that

$$\sup_{t\geq -r} \int_t^{t+1} |g(\tau)|\, d\tau < +\infty,$$

identified if they are equal almost everywhere. The following result, also obtained in [12], is established in Sec. 5.5.

Theorem 1.5. *Equation* (1.4) *has an exponential dichotomy on* $\mathbb{R}_0^+$ *if and only if there exists a closed subspace* $B \subset C$ *such that for each* $g \in \mathcal{F}^+$ *there exists a unique* $v \in \mathcal{E}_B^+$ *such that* (1.15) *holds for* $t \geq s \geq 0$.

The space B in Theorem 1.5 turns out to be the unstable space at time 0.

To the possible extent, the proofs of Theorems 1.4 and 1.5 follow the approach developed in [1] in the context of ordinary differential equations on Banach spaces, although it requires considerable modifications due to the context of delay equations. The latter approach was then extended to delay equations in [12], on which our presentation is inspired.

In Chap. 6 we use the characterizations of an exponential dichotomy in terms of an admissibility property in Theorems 1.4 and 1.5 to establish the robustness of the notion under sufficiently small perturbations (see Sec. 1.2 for details and references on the robustness problem). In particular, we give a second proof of Theorem 1.2 (and in fact of Theorem 4.1), now using admissibility. We also give a proof of the robustness of an exponential dichotomy on the half-line (Chap. 4 considers only exponential dichotomies on the line, with a method that cannot be used on the half-line).

Finally, also in Chap. 6, we consider perturbations depending on a parameter and we obtain parameterized versions of the robustness property for Lipschitz and smooth perturbations. Johnson and Sell [65] considered exponential dichotomies on $\mathbb{R}$ and for C^k perturbations showed that if the perturbation and its derivatives are bounded and equicontinuous (in the parameter), then the projections are of class C^k. Palmer [102] considered the same problem for exponential dichotomies on $\mathbb{R}_0^+$ and showed that by fixing the null space the corresponding projections are of class $C^{1+\text{Lip}}$ provided that the perturbation has this regularity. In the case of discrete time, the robustness under smooth parameterized perturbations of exponential dichotomies was first established in [7].

1.4   Smooth Invariant Manifolds

Stable Manifolds: Lipschitz Perturbations

We also discuss the behavior of the solutions of a linear delay equation under sufficiently small nonlinear perturbations, first for Lipschitz perturbations and then for smooth ones. We start by constructing Lipschitz stable and unstable invariant manifolds for any sufficiently small Lipschitz perturbation of an exponential dichotomy in Chap. 7. More precisely, we show that the set of initial conditions leading to a bounded forward global solution is a graph of a Lipschitz function over the stable bundle, which is precisely the stable manifold. A similar statement holds for the bounded backward global solutions, leading to the construction of the unstable manifold. In the latter case one also needs to show that the solutions exist and are unique, besides being global (since they need to go backwards in time).

We formulate briefly the main results established in Chap. 7. We start with the construction of a Lipschitz stable invariant manifold for the delay equation

$$v' = L(t)v_t + g(t, v_t) \tag{1.16}$$

assuming that the linear Eq. (1.4) has an exponential dichotomy on $\mathbb{R}_0^+$ and that g is a sufficiently small Lipschitz perturbation (in fact we consider the more general case of an exponential dichotomy on an interval $I \subset \mathbb{R}$ containing $\mathbb{R}_0^+$). In addition, we assume that:

(1) $L(t) \colon C \to \mathbb{R}^n$, for $t \geq 0$, are bounded linear operators for which the map $(t, v) \mapsto L(t)v$ is continuous on $\mathbb{R}_0^+ \times C$ and condition (1.9) holds;
(2) $g \colon \mathbb{R}_0^+ \times C \to \mathbb{R}^n$ is continuous and $g(t, 0) = 0$ for all $t \geq 0$.

The *stable set* V^s of Eq. (1.16) is the set of all initial conditions $(s, \varphi) \in \mathbb{R}_0^+ \times C$ for which the solution $v = v(\cdot, s, \varphi)$ with $v_s = \varphi$ is defined and bounded on the interval $[s-r, +\infty)$. We show in particular that if Eq. (1.4) has an exponential dichotomy on $\mathbb{R}_0^+$, then its stable set is composed of the pairs $(s, \varphi) \in \mathbb{R}_0^+ \times C$ for which there exists a bounded continuous function $v \colon [s - r, +\infty) \to \mathbb{R}^n$ such that $v_s = \varphi$ and

$$v_t = T(t, s)P(s)\varphi + \int_s^t T_0(t, \tau)P_0(\tau)g(\tau, v_\tau)\,d\tau$$
$$- \int_t^{+\infty} \overline{T}(t, \tau)Q_0(\tau)g(\tau, v_\tau)\,d\tau$$

for every $t \geq s$ (see Proposition 7.2). Using this characterization of the stable set V^s, one can show that all bounded solutions of Eq. (1.16) decay exponentially (see Proposition 7.3).

Now we formulate the Lipschitz stable manifold theorem. We denote by $E(s)$ and $F(s)$, respectively, the stable and unstable spaces associated with the exponential dichotomy. Let $\mathcal{S}^L$ be the set of all continuous functions

$$z \colon \bigl\{(s, a) \in \mathbb{R}_0^+ \times C : a \in E(s)\bigr\} \to C$$

such that for each $s \geq 0$:

(1) $z(s, 0) = 0$ and $z(s, E(s)) \subset F(s)$;
(2) for $a, \overline{a} \in E(s)$ we have

$$\|z(s, a) - z(s, \overline{a})\| \leq \|a - \overline{a}\|.$$

For each function $z \in \mathcal{S}^L$ we consider its graph

$$\text{graph}\, z = \big\{(s, a + z(s,a)) : (s,a) \in \mathbb{R}_0^+ \times E(s)\big\} \subset \mathbb{R}_0^+ \times C.$$

The following result is a particular case of Theorem 7.3.

Theorem 1.6. *Assume that Eq. (1.4) has an exponential dichotomy on $\mathbb{R}_0^+$ and that there exists $\delta > 0$ such that*

$$|g(t,u) - g(t,v)| \le \delta \|u - v\|$$

for every $t \ge 0$ and $u, v \in C$. Then, for any sufficiently small δ, there exists a function $z \in \mathcal{S}^L$ such that $V^s = \text{graph}\, z$.

Theorem 1.6 is due to Hale and Perelló [53]. See Chap. 7 for the construction of Lipschitz unstable manifolds and for the study of the Lipschitz dependence of the manifolds on a parameter assuming that the perturbation has this behavior.

We also establish a partial version of the Grobman–Hartman theorem in the theory of ordinary differential equations for Lipschitz perturbations of an exponential dichotomy on $\mathbb{R}$. Let $\mathcal{M}$ be the set of all continuous functions

$$\eta \colon \big\{(t,b) : t \in \mathbb{R}, b \in F(t)\big\} \to C$$

such that

$$\|\eta\|_\infty := \sup\big\{\|\eta(t,b)\| : t \in \mathbb{R}, b \in F(t)\big\} < +\infty.$$

Moreover, let

$$\eta^t = \eta(t, \cdot) \quad \text{and} \quad h^t = \text{Id}_{F(t)} + \eta^t.$$

We continue to denote by $T(t,s)$ the evolution family associated with the linear equation $v' = L(t)v_t$ and we write the solutions of Eq. (1.16) in the form

$$v_t = R(t,s)(v_s) \quad \text{for } t \ge s.$$

It is shown in Theorem 7.6 that if

$$|g(t,u) - g(t,v)| \le \delta \min\{1, \|u - v\|\}$$

for every $t \in \mathbb{R}$ and $u, v \in C$, and some sufficiently small $\delta > 0$, then there exists a unique $\eta \in \mathcal{M}$ such that

$$h^t \circ T(t,s) = R(t,s) \circ h^s \quad \text{on } F(s)$$

for every $t, s \in \mathbb{R}$ with $t \ge s$. Moreover, each map h^t is one-to-one.

In the special case of autonomous delay equations, a somewhat related result is due to Farkas [37], although he only obtained a local conjugacy, either for a fixed time or at most for sufficiently small times. Moreover, our proof follows simpler direct arguments in [6] that avoid discrete time. N. Sternberg [129, 130] gave earlier extensions of the Grobman–Hartman theorem for delay equations under the restrictive assumption of the existence of a global attractor.

The original references for the Grobman–Hartman theorem for ordinary differential equations are Grobman [41, 42] and Hartman [55, 56]. Using the ideas in Moser's proof in [94] of the structural stability of Anosov diffeomorphisms, the Grobman–Hartman theorem was extended to Banach spaces independently by Palis [100] and Pugh [117]. The work of S. Sternberg [131, 132] showed that there are algebraic obstructions, expressed in terms of resonances in the autonomous case, that prevent the existence of conjugacies with a prescribed high regularity (see also [13, 14, 89, 124] for further related work).

Stable Manifolds: Smooth Perturbations

In Chap. 8 we construct smooth stable and unstable invariant manifolds for any sufficiently small smooth perturbation of an exponential dichotomy. In view of the uniqueness of the Lipschitz invariant manifolds already constructed in Chap. 7 it remains to show that the function of which the invariant manifold is a graph has the required regularity properties.

We also formulate briefly the main results in Chap. 8. We consider again nonlinear perturbations of Eq. (1.4) and we establish the existence of a smooth stable invariant manifold for Eq. (1.16) when the linear equation $v' = L(t)v_t$ has an exponential dichotomy on an interval $I \subset \mathbb{R}$ containing $\mathbb{R}_0^+$ and g is a sufficiently small C^1 perturbation. We assume that:

(1) $L(t) \colon C \to \mathbb{R}^n$, for $t \in I$, are bounded linear operators for which the map $(t, v) \mapsto L(t)v$ is of class C^1 and condition (1.9) holds;
(2) $g \colon I \times C \to \mathbb{R}^n$ is of class C^1, and $g(t, 0) = 0$ and $\partial g(t, 0) = 0$ for all $t \in I$, where ∂ denotes the partial derivative with respect to the second variable.

The stable manifold is now obtained as a graph of a function of class C^1 in the second variable. Let $\mathcal{S}^1$ be the set of all functions $z \in \mathcal{S}^L$ of class C^1 in the second variable such that $\partial z(s, 0) = 0$ for $s \in I$. The following result is the content of Theorem 8.1.

Theorem 1.7. *Assume that Eq. (1.4) has an exponential dichotomy on an interval I containing $\mathbb{R}_0^+$ and that there exists $\delta > 0$ such that $\|\partial g(t,v)\| \leq \delta$ for every $t \in I$ and $v \in C$. Then, for any sufficiently small δ, the function $z \in \mathcal{S}^L$ given by Theorem 1.6 is in $\mathcal{S}^1$.*

We also consider the case when g is a sufficiently small C^1 perturbation with Lipschitz derivative. In comparison to Theorem 1.7, we show, in addition, that for the unique function $z \in \mathcal{S}^L$ given by Theorem 1.6 each function $s \mapsto z(s,a)$ has a Lipschitz derivative. We emphasize that both Theorem 1.7 and the present result establish the optimal regularity of the stable manifolds.

Assume that Eq. (1.4) has an exponential dichotomy on an interval $I \subset \mathbb{R}$ containing $\mathbb{R}_0^+$. Let $\mathcal{S}^{1+L}$ be the set of all functions $z \in \mathcal{S}^1$ such that

$$\|\partial z(s,a) - \partial z(s,\overline{a})\| \leq \|a - \overline{a}\|$$

for every $s \in I$ and $a, \overline{a} \in E(s)$. Note that a function $z \in \mathcal{S}^L$ is in $\mathcal{S}^{1+L}$ if and only if it is of class C^1 in a and for each $s \in I$:

(1) $z(s,0) = 0$, $\partial z(s,0) = 0$ and $z(s, E(s)) \subset F(s)$;
(2) for $a, \overline{a} \in E(s)$ we have

$$\|\partial z(s,a)\| \leq 1 \quad \text{and} \quad \|\partial z(s,a) - \partial z(s,\overline{a})\| \leq \|a - \overline{a}\|.$$

Theorem 1.8. *Assume that Eq. (1.4) has an exponential dichotomy on an interval I containing $\mathbb{R}_0^+$ and that there exists $\delta > 0$ such that*

$$\|\partial g(t,u)\| \leq \delta \quad \text{and} \quad \|\partial g(t,u) - \partial g(t,v)\| \leq \delta\|u - v\|$$

for every $t \in I$ and $u, v \in C$. Then, for any sufficiently small δ, the function $z \in \mathcal{S}^L$ given by Theorem 1.6 is in $\mathcal{S}^{1+L}$.

We refer to Chap. 8 for the construction of stable and unstable invariant manifolds with higher smoothness.

Center Manifolds

In Chap. 9 we turn to center manifolds. Our main goal is to give a streamlined detailed proof of the existence of smooth center invariant manifolds for any sufficiently small nonautonomous perturbation of a linear delay equation with an exponential trichotomy. More precisely, we consider C^k perturbations with Lipschitz kth derivative and we obtain the same regularity for the center manifolds.

Consider the linear delay Eq. (1.4), where $L(t)\colon C \to \mathbb{R}^n$, for $t \in \mathbb{R}$, are bounded linear operators. We assume that the map $(t, \varphi) \to L(t)\varphi$ is of class C^k on $\mathbb{R} \times C$ and that

$$\sup_{t \in \mathbb{R}} \int_t^{t+1} \|L(\tau)\|\, d\tau < +\infty.$$

We say that Eq. (1.4) has an *exponential trichotomy* (on $\mathbb{R}$) if:

(1) there exist projections $P(t), Q(t), R(t)\colon C \to C$, for $t \in \mathbb{R}$, satisfying

$$P(t) + Q(t) + R(t) = \mathrm{Id},$$

such that for $t \geq s$ we have

$$P(t)T(t,s) = T(t,s)P(s), \quad Q(t)T(t,s) = T(t,s)Q(s)$$

and

$$R(t)T(t,s) = T(t,s)R(s);$$

(2) the linear operator

$$\overline{T}(t,s) = T(t,s)|\operatorname{Ker} P(s)\colon \operatorname{Ker} P(s) \to \operatorname{Ker} P(t)$$

is invertible for $t \geq s$;

(3) there exist constants $\lambda, \mu, D > 0$ with $\mu < \lambda$ such that

$$\|T(t,s)R(s)\| \leq De^{\mu(t-s)}, \quad \|T(t,s)P(s)\| \leq De^{-\lambda(t-s)}$$

and

$$\|\overline{T}(s,t)R(t)\| \leq De^{\mu(t-s)}, \quad \|\overline{T}(s,t)Q(t)\| \leq De^{-\lambda(t-s)}$$

for $t \geq s$, where $\overline{T}(s,t) = \overline{T}(t,s)^{-1}$.

Now we consider the perturbed Eq. (1.16), where $g\colon \mathbb{R} \times C \to \mathbb{R}^n$ is a C^k map with $g(t,0) = 0$ and $\partial g(t,0) = 0$ for all $t \in \mathbb{R}$. The *center set* V^c of Eq. (1.16) is the set of all initial conditions $(s, \varphi) \in \mathbb{R} \times C$ for which there exists a solution $v\colon \mathbb{R} \to \mathbb{R}^n$ of the equation with $v_s = \varphi$ satisfying

$$\sup_{t \in \mathbb{R}} \left(\|v_t\| e^{-\nu|t|} \right) < +\infty$$

for any sufficiently large $\nu < \lambda$.

The following result is a combination of Theorems 9.2 and 9.3 that were obtained in [11]. Let

$$E(t) = P(t)C, \quad F(t) = Q(t)C \quad \text{and} \quad G(t) = R(t)C.$$

Theorem 1.9. *Assume that Eq. (1.4) has an exponential trichotomy and that there exists $\delta > 0$ such that*

$$\|\partial^j g(t, u)\| \leq \delta \quad \text{and} \quad \|\partial^k g(t, u) - \partial^k g(t, v)\| \leq \delta \|u - v\|$$

for every $t \in \mathbb{R}$, $u, v \in C$ and $j = 1, \ldots, k$. If $\lambda > (k + 1)\mu$ and δ is sufficiently small, then there exists a function

$$z \colon \big\{ (s, c) \in \mathbb{R} \times C : c \in G(s) \big\} \to C$$

of class C^k in c with $V^c = \operatorname{graph} z$ such that for each $s \in \mathbb{R}$:

(1) $z(s, 0) = 0$, $\partial z(s, 0) = 0$ and $z(s, G(s)) \subset E(s) \oplus F(s)$;
(2) for $j = 1, \ldots, k$ and $c, d \in G(s)$ we have

$$\|\partial^j z(s, c)\| \leq 1 \quad \text{and} \quad \|\partial^k z(s, c) - \partial^k z(s, d)\| \leq \|c - d\|.$$

Again we obtain V^c as a graph after solving certain fixed point problems that must be satisfied by the global bounded solutions of Eq. (1.16). To obtain the bounds that are required for the fixed point problems we use a multivariate version of the Faà di Bruno formula for the derivatives of a composition (see [29]). We also use a result of Henry in [58] that allows one to establish the existence and regularity of the center manifolds using a single fixed point problem, instead of one for each derivative up to order k. His result says that a closed ball in the space $C^{k+\mathrm{Lip}}$ of all functions of class C^k with Lipschitz kth derivative (with a given Lipschitz constant) is closed with respect to the C^0 topology. This approach was developed in [11], on which our presentation is inspired.

A sketch of an alternative proof of Theorem 1.9 is given in [54]. For a detailed proof of the center manifold theorem using the sun-star machinery see [34]. Some early works considering various types of equations and perturbations are due to Chafee [20, 21], Kurzweil [71, 72] and Fodcuk [38, 39].

Center manifolds are powerful tools in the analysis of the behavior of a dynamical system. For example, when Eq. (1.4) has no unstable directions, all solutions converge exponentially to the center manifold. Therefore, the stability of the system is determined by the behavior on the center manifold (we refer to the book [19] for details and references). This causes that it is often sufficient to consider a reduction of the dynamics to the center manifolds. Moreover, this has also the advantage of reducing the dimension of the system, which in the case of delay equations is quite substantial since for $r > 0$ the unstable spaces are finite-dimensional. In particular, using normal forms has important consequences for the study of bifurcations [35, 36] (see also [40, 44, 45, 66, 80, 97, 98]). In these studies one needs to approximate the center manifolds to sufficiently high order, which causes that it is also important to discuss the smoothness of the manifolds. The classical theory of center manifolds goes back to Pliss [112] and Kelley [67], with further developments due to Carr [19], Hirsch, Pugh and Shub [62], Sijbrand [128], Guckenheimer and Holmes [43] and others. For example, in the autonomous case the optimal regularity of the center (unstable) manifolds for C^k perturbations with Lipschitz kth derivative was established in [27]. A detailed exposition of the theory in the case of autonomous equations (in the context of ordinary differential equations) is given in [135], adapting results in [137]. See also [22,90,136] for the infinite-dimensional case.

In another direction, the properties of reversibility and equivariance of a dynamical system descend to any center manifold. More precisely, the dynamics on any center manifold of a reversible (respectively equivariant) system is also reversible (respectively equivariant). It should be noted that time-reversal symmetries are fundamental symmetries in many physical systems, both in classical and quantum mechanics. This is due to the fact that many Hamiltonian systems are reversible (see [73,118] for many examples). See [90,126] and the references therein for the discussion of the consequences of reversibility.

1.5 Spectra and Perturbations

In Chap. 10 we consider the spectrum of a nonautonomous linear delay equation and we characterize all its possible forms. This spectrum is inspired on the one introduced by Sacker and Sell in [121] for linear cocycles over a semiflow (or, equivalently, for linear skew-product semiflows). A related spectrum for delay equations was first considered by Magalhães in [83]. For further work we refer the reader to [23–25,64,76,122] and the references therein.

Consider the linear delay equation

$$v' = L(t)v_t$$

and its evolution family $T(t, s)$. The *spectrum* of the equation is the set Σ of all numbers $a \in \mathbb{R}$ such that the evolution family

$$T_a(t, s) = e^{-a(t-s)}T(t, s)$$

does not have an exponential dichotomy on $\mathbb{R}$. The following theorem describes all possible forms of the spectrum for a linear delay equation (see Theorem 10.1). Given $-\infty \leq a \leq b \leq +\infty$, we write $|a, b| = \mathbb{R} \cap [a, b]$.

Theorem 1.10. *Assume that $r > 0$. For the spectrum of Eq. (1.4), one of the following alternatives holds:*

(1) $\Sigma = \emptyset$;

(2) $\Sigma = \bigcup_{m=1}^{k} |a_m, b_m|$ *for some numbers*

$$+\infty \geq b_1 \geq a_1 > b_2 \geq a_2 > \cdots > b_k \geq a_k \geq -\infty$$

with $k \in \mathbb{N}$ (when $k = 1$, $a_1 = -\infty$ and $b_1 = +\infty$ we have $\Sigma = \mathbb{R}$);

(3) $\Sigma = \bigcup_{m=1}^{\infty} |a_m, b_m|$ *for some numbers*

$$+\infty \geq b_1 \geq a_1 > b_2 \geq a_2 > b_3 \geq a_3 > \cdots \qquad (1.17)$$

with $a_m \to -\infty$ when $m \to \infty$;

(4) $\Sigma = \bigcup_{m=1}^{\infty} |a_m, b_m| \cup (-\infty, a_\infty]$ *for some numbers as in (1.17) with $a_m \to a_\infty \in \mathbb{R}$ when $m \to \infty$.*

To the possible extent, our proof of Theorem 10.1 follows [9], which considers the case of ordinary differential equations. For simplicity of the exposition, in the remainder of the section we assume that Σ is given by alternative 2 in Theorem 10.1 (see Sec. 10.2 for the general case).

Now we associate spaces $G_I(s)$, for $s \in \mathbb{R}$, to each connected component $I \subset \Sigma$. It turns out that the Lyapunov exponents of the initial conditions in $G_I(s)$ belong to I. For each $a \in \mathbb{R}$ and $t \in \mathbb{R}$, let

$$E_a(s) = \left\{ \varphi \in C : \sup_{t \geq s}\big(e^{-a(t-s)}\|T(t,s)\varphi\|\big) < +\infty \right\}$$

and let $F_a(s)$ be the set of all $\varphi \in C$ for which there exists a function $v: (-\infty, s] \to \mathbb{R}^n$ with $v_s = \varphi$ such that $v_t = T(t,\tau)v_\tau$ for $s \geq t \geq \tau$ and

$$\sup_{t \leq s}\big(e^{-a(t-s)}|v(t)|\big) < +\infty.$$

For each connected component $I = |a_m, b_m|$, with $m = 1, \ldots, k$, we define

$$G_I(s) = E_{c_{m-1}}(s) \cap F_{c_m}(s),$$

where $c_m \in (b_{m+1}, a_m)$ for $m = 1, \ldots, k - 1$,

$$c_0 \geq b_1 + \delta \quad \text{and} \quad c_k \leq b_m - \delta$$

with $\delta > 0$. Note that $c_0 = +\infty$ when $b_1 = +\infty$ and that $c_k = -\infty$ when $b_m = -\infty$. One can show that the spaces $G_I(s)$ are independent of the choice of numbers c_m and δ. The following result shows that the Lyapunov exponents of the initial conditions in the space $G_I(s)$ belong to I (this is the content of Theorem 10.2).

Theorem 1.11. *Assume that $r > 0$. For each $s \in \mathbb{R}$ and $\varphi \in G_I(s) \setminus \{0\}$, we have*

$$\left(\liminf_{t \to +\infty} \frac{1}{t} \log\|T(t,s)\varphi\|, \limsup_{t \to +\infty} \frac{1}{t} \log\|T(t,s)\varphi\| \right) \subset I$$

and there exists a function $v \colon (-\infty, s] \to \mathbb{R}^n$ with $v_s = \varphi$ such that

$$v_\tau = T(\tau, \sigma)v_\sigma \quad \text{for } s \geq \tau \geq \sigma$$

and

$$\left(\liminf_{t \to -\infty} \frac{1}{t} \log\|v_t\|, \limsup_{t \to -\infty} \frac{1}{t} \log\|v_t\| \right) \subset I.$$

See Sec. 10.3 for corresponding results when the linear operators $L(t)$ are defined only for $t \geq 0$.

We also show that the asymptotic behavior described in Theorem 1.11 persists under sufficiently small perturbations. Consider the equation

$$v' = L(t)v_t + g(t, v_t), \tag{1.18}$$

where $g \colon \mathbb{R} \times C \to \mathbb{R}^n$ is a measurable function.

Theorem 1.12. *Assume that $r > 0$ and that condition (1.9) holds. Let $v \colon [s - r, +\infty) \to \mathbb{R}^n$ be a solution of Eq. (1.18) satisfying*

$$\lim_{t \to +\infty} \int_t^{t+1} \frac{|g(\tau, v_\tau)|}{\|v_\tau\|} \, d\tau = 0.$$

If there exists $p \in \mathbb{N}$ for which

$$\liminf_{t \to +\infty} \frac{1}{t} \log\|v_t\| \geq a_p,$$

then there exists $i \in \{1, \ldots, p\}$ such that

$$\left(\liminf_{t \to +\infty} \frac{1}{t} \log\|v_t\|, \limsup_{t \to +\infty} \frac{1}{t} \log\|v_t\| \right) \subset (a_i, b_i).$$

Our proof of Theorem 1.12 follows [8], which considers ordinary differential equations. In the particular case of perturbations of an autonomous linear equation with constant coefficients $x' = Ax$ (for which the Lyapunov exponents are always limits), a related result can be found in Coppel's book [30]. Earlier results were obtained by Perron [104], Lettenmeyer [77] and Hartman and Wintner [57]. Corresponding results for perturbations of autonomous delay equations $v' = Lv_t$ were established by Pituk [110, 111] (for finite delay) and by Matsui, Matsunaga and Murakami [88] (for infinite delay). Related results for perturbations of autonomous difference equations were first obtained by Coffman [28].

Chapter 2

Basic Notions

In this chapter we present various basic notions and results from the theory of delay equations that are used throughout the book. In particular, we discuss the existence, uniqueness and continuation of solutions under the Carathéodory conditions. After this introduction to the theory, we consider a particular class of nonautonomous linear delay equations. We introduce their associated evolution families and we discuss how these can be extended to certain discontinuous functions. Both developments are crucial for the formulation of the variation of constants formula and its proof in this chapter, as well as for many developments in other chapters. Finally, we consider the simpler class of autonomous linear delay equations and their nonautonomous perturbations.

2.1 Delay Equations: Basic Results

In this section we present some basic notions and results from the general theory of delay equations.

Let $|\cdot|$ be the norm on $\mathbb{R}^n$. Given $r \geq 0$, we denote by C the Banach space of all continuous functions $\varphi \colon [-r, 0] \to \mathbb{R}^n$ with the norm

$$\|\varphi\| = \sup\{|\varphi(\theta)| : -r \leq \theta \leq 0\}. \tag{2.1}$$

Definition 2.1. We say that a function $f \colon \Omega \to \mathbb{R}^n$ satisfies the *Carathéodory conditions* on an open set $\Omega \subset \mathbb{R} \times C$ if:

(1) $t \mapsto f(t, \varphi)$ is measurable for each φ;
(2) $\varphi \mapsto f(t, \varphi)$ is continuous for almost all t;
(3) given $(s, \varphi) \in \Omega$, there exists an open neighborhood $U \subset \Omega$ of (s, φ) and an integrable function $h \colon \mathbb{R} \to \mathbb{R}^+$ such that

$$|f(t, \psi)| \leq h(t) \quad \text{for } (t, \psi) \in U.$$

Note that any continuous f satisfies the Carathéodory conditions.

Assume that $f\colon \Omega \to \mathbb{R}^n$ satisfies the Carathéodory conditions on an open set $\Omega \subset \mathbb{R} \times C$. We consider the *delay equation*

$$v' = f(t, v_t), \qquad (2.2)$$

where v' denotes the right-hand derivative (when it exists) and

$$v_t(\theta) = v(t + \theta) \quad \text{for } \theta \in [-r, 0]$$

(assuming that v is defined at least on the interval $[t - r, t]$).

We observe that delay equations include ordinary differential equations as a (very) particular case. Indeed, for $r = 0$, the spaces C and $\mathbb{R}^n$ coincide (including their norms, taking $r = 0$ in (2.1)) and one can define a function g on some subset of $\mathbb{R} \times \mathbb{R}^n$ by

$$g(t, \varphi(0)) = f(t, \varphi)$$

for $(t, \varphi) \in \Omega$. Hence, in this case Eq. (2.2) is simply the ordinary differential equation

$$v' = g(t, v).$$

On the other hand, the class of delay equations is much larger than the class of ordinary differential equations. Various similarities and various differences between the properties of the two classes shall be observed along the book. However, the space C is infinite-dimensional when $r > 0$, which causes already some unavoidable complications.

Now we introduce the notion of a solution.

Definition 2.2. A continuous function $v\colon [s - r, a) \to \mathbb{R}^n$ with $a \le +\infty$ is called a *solution* of Eq. (2.2) if $(t, v_t) \in \Omega$ and

$$v(t) = v(s) + \int_s^t f(\tau, v_\tau)\, d\tau \quad \text{for } t \in [s, a). \qquad (2.3)$$

Since v is uniformly continuous on bounded intervals, the map $t \mapsto v_t$ is continuous and so it follows from condition 3 in Definition 2.1 that the integral in (2.3) is well defined for any sufficiently small a. One can show that $v\colon [s - r, a) \to \mathbb{R}^n$ is a solution of Eq. (2.2) if and only if v is absolutely continuous on $[s, a)$ and $v' = f(t, v_t)$ for almost every $t \in [s, a)$. When f is continuous, the map $t \mapsto f(t, v_t)$ is continuous on $[s, a)$ and so it follows from (2.3) that v is of class C^1 on $[s, a)$ and $v' = f(t, v_t)$ for all $t \in [s, a)$.

One can also consider solutions with domain $(-\infty, a)$.

Definition 2.3. A continuous function $v\colon (-\infty, a) \to \mathbb{R}^n$ with $a \le +\infty$ is called a *solution* of Eq. (2.2) if the restriction of v to $[s - r, a)$ is a solution of the equation for all $s \in (-\infty, a)$.

Finally, we introduce the notion of initial value problem.

Definition 2.4. Given $(s, \varphi) \in \Omega$, the *initial value problem*

$$\begin{cases} v' = f(t, v_t), \\ v_s = \varphi \end{cases} \tag{2.4}$$

consists of finding a solution $v \colon [s - r, a) \to \mathbb{R}^n$ of Eq. (2.2), for some $a > s$, satisfying the *initial condition* $v_s = \varphi$.

We start by discussing the existence of solutions.

Theorem 2.1 (Existence). *Given a function $f \colon \Omega \to \mathbb{R}^n$ satisfying the Carathéodory conditions on an open set $\Omega \subset \mathbb{R} \times C$ and $(s, \varphi) \in \Omega$, there exists a solution of the initial value problem (2.4).*

Proof. Since f satisfies the Carathéodory conditions, there exist constants $\alpha, \beta > 0$ such that

$$\int_s^{s+\alpha} h(\tau) \, d\tau \leq \beta$$

and

$$[s, s + \alpha] \times \{ \psi \in C : \|\psi - \varphi\| \leq \beta \} \subset U,$$

with U as in Definition 2.1. Given $m \in \mathbb{N}$, let

$$I_j = (s + j\alpha/m, s + (j+1)\alpha/m]$$

for $j = 0, \ldots, m - 1$ and define a function $c_m \colon [s, s + \alpha] \to [s, s + \alpha)$ by

$$c_m(t) = \begin{cases} s + j\alpha/m & \text{if } t \in I_j \text{ for some } j = 0, \ldots, m - 1, \\ t & \text{if } t \in [s - r, s]. \end{cases}$$

We also define $v^m \colon [s - r, s + \alpha] \to \mathbb{R}^n$ by $v_s^m = \varphi$ and recursively by

$$v^m(t) = v^m(s + j\alpha/m) + \int_{s+j\alpha/m}^t f(\tau, v_{c_m(\tau)}^m) \, d\tau$$

for $t \in I_j$ and $j = 0, \ldots, m - 1$, where

$$v_{c_m(\tau)}^m(\theta) = v^m(c_m(\tau) + \theta).$$

Then

$$v^m(t) = \varphi(0) + \int_s^t f(\tau, v_{c_m(\tau)}^m) \, d\tau. \tag{2.5}$$

We have

$$|v^m(t) - \varphi(0)| \le \int_s^t h(\tau)\,d\tau \le \beta$$

and so the sequence $(v^m)_{m \in \mathbb{N}}$ is bounded. Moreover, for each $t, \bar{t} \in [s, s+\alpha]$, we obtain

$$|v^m(t) - v^m(\bar{t})| \le \left| \int_t^{\bar{t}} f(\tau, v^m_{c_m(\tau)})\,d\tau \right|$$
$$\le \left| \int_t^{\bar{t}} h(\tau)\,d\tau \right| \le |F(t) - F(\bar{t})|, \tag{2.6}$$

where

$$F(t) = \int_s^t h(\tau)\,d\tau.$$

Since F is continuous (and thus uniformly continuous on $[s, s+\alpha]$), the sequence $(v^m)_{m \in \mathbb{N}}$ is equicontinuous. By the Arzelà–Ascoli theorem, it has a subsequence $(v^{p_m})_{m \in \mathbb{N}}$ converging uniformly to a continuous function $v \colon [s-r, s+\alpha] \to \mathbb{R}^n$ on $[s-r, s+\alpha]$. Let $w^m = v^{p_m}$ and $d_m = c_{p_m}$. Given $\varepsilon > 0$, since $|c_m(\tau) - \tau| < 1/m$, there exists $m \in \mathbb{N}$ such that

$$\sup_{\tau \in [s,s+\alpha], \theta \in [-r, 0]} |w^m(d_m(\tau) + \theta) - w^m(\tau + \theta)| < \varepsilon$$

and

$$\sup_{\tau \in [s-r, s+\alpha]} |w^m(\tau) - v(\tau)| < \varepsilon.$$

Thus,

$$\|w^m_{d_m(\tau)} - v_\tau\| \le \|w^m_{d_m(\tau)} - w^m_\tau\| + \|w^m_\tau - v_\tau\| < 2\varepsilon$$

and so

$$w^m_{d_m(\tau)} \to v_\tau \quad \text{when } m \to \infty \tag{2.7}$$

(uniformly for $\tau \in [s-r, s+\alpha]$). Now we consider the functions

$$f_m(\tau) = f(\tau, w^m_{d_m(\tau)}).$$

We have $|f_m(\tau)| \le h(\tau)$ for all $m \in \mathbb{N}$ and $\tau \in [s, s+\alpha]$. Moreover, by (2.7), it follows from condition 2 in Definition 2.1 that $f_m(\tau) \to f(\tau, v_\tau)$ when $m \to \infty$, for almost all $\tau \in [s, s+\alpha]$. Therefore, by the dominated convergence theorem, it follows from (2.5) that

$$v(t) = \varphi(0) + \int_s^t f(\tau, v_\tau)\,d\tau$$

for $t \in [s, s+\alpha]$. Moreover, $v_s = \varphi$. Hence, v is a solution of the initial value problem (2.4) on the interval $[s-r, s+\alpha]$. $\qquad\square$

The uniqueness of solutions can be established for example under the following assumption. A function $f\colon \Omega \to \mathbb{R}^n$ is said to be *locally Lipschitz* if for each compact set $V \subset \Omega$ there exists $\delta > 0$ such that

$$|f(t,\varphi) - f(t,\psi)| \le \delta \|\varphi - \psi\| \tag{2.8}$$

for every $(t,\varphi), (t,\psi) \in V$. Note that any C^1 function is locally Lipschitz.

Theorem 2.2 (Uniqueness). *Let $f\colon \Omega \to \mathbb{R}^n$ be a locally Lipschitz function satisfying the Carathéodory conditions on an open set $\Omega \subset \mathbb{R} \times C$. Then for each $(s,\varphi) \in \Omega$ the initial value problem (2.4) has a unique solution on some interval $[s-r, a)$.*

Proof. Let v and w be solutions of the initial value problem (2.4) on the same interval $[s-r,a)$ (the existence of at least one solution is guaranteed by Theorem 2.1). Then

$$v(t) = v(s) + \int_s^t f(\tau, v_\tau) \, d\tau$$

and

$$w(t) = w(s) + \int_s^t f(\tau, w_\tau) \, d\tau$$

for $t \in [s,a)$. Since $v_s = w_s = \varphi$, we obtain

$$v(t) - w(t) = \int_s^t \left(f(\tau, v_\tau) - f(\tau, w_\tau) \right) d\tau$$

for $t \in [s,a)$. Hence,

$$|v(t+\theta) - w(t+\theta)| \le \int_s^{t+\theta} \|f(\tau, v_\tau) - f(\tau, w_\tau)\| \, d\tau$$

$$\le \int_s^t \|f(\tau, v_\tau) - f(\tau, w_\tau)\| \, d\tau$$

whenever $t + \theta \in [s,a)$ with $\theta \in [-r, 0]$. Therefore,

$$\|v_t - w_t\| \le \int_s^t \|f(\tau, v_\tau) - f(\tau, w_\tau)\| \, d\tau \tag{2.9}$$

for $t \in [s,a)$. In order to apply property (2.8) we must show that there exists a compact set $V \subset \Omega$ satisfying

$$(t, v_t), (t, w_t) \in V \text{ for } t \in [s,a). \tag{2.10}$$

Proceeding as in (2.6) we obtain

$$|v(t) - v(\bar{t})| = \left| \int_t^{\bar{t}} f(\tau, v_\tau) \, d\tau \right| \le |F(t) - F(\bar{t})|,$$

where $F(t) = \int_s^t h(\tau)\, d\tau$. Hence, v is uniformly continuous on $[s - r, a)$, which implies that

$$A = \big\{(t, v_t) : t \in [s, a)\big\}$$

is contained in a compact subset of Ω. Indeed, take a sequence (t_m, v_{t_m}) in A and let s_m be a converging subsequence of t_m. If $s_m \to b < a$, then $v_{s_m} \to v_b$ uniformly (this follows readily from the uniform continuity of v). On the other hand, if $s_m \to a$, then $v_{s_m} \to \psi$, where

$$\psi(\theta) = \begin{cases} v_a(\theta) & \text{if } \theta \in [-r, 0), \\ p & \text{if } \theta = 0 \end{cases}$$

and

$$p = \lim_{m \to \infty} v(s_m) = \lim_{m \to \infty} v(t_m)$$

(the limits exist and are independent of the sequence t_m, also in view of the uniform continuity of v). Hence, A is contained in the compact set $A \cup \{(a, \psi)\}$. A similar argument applies to the set $\{(t, w_t) : t \in [s, a)\}$. Therefore, there exists a compact set $V \subset \Omega$ satisfying (2.10) and it follows from (2.8) and (2.9) that

$$\|v_t - w_t\| \le \delta \int_s^t \|v_\tau - w_\tau\|\, d\tau$$

for $t \in [s, a)$. Applying Gronwall's lemma, we conclude that $v_t = w_t$ for $t \in [s, a)$ and so $v = w$ on $[s - r, a)$ (because $v_s = w_s = \varphi$). $\qquad \square$

We also discuss the continuation (or extension) of solutions. A solution v of Eq. (2.2) on $[s-r, a)$ is said to be *noncontinuable* if there exists no solution w of Eq. (2.2) on some interval $[s - r, b)$, with $b > a$, such that $w = v$ on $[s - r, a)$. Otherwise the solution v is said to be *continuable*.

Theorem 2.3 (Continuation). *Given a function $f \colon \Omega \to \mathbb{R}^n$ satisfying the Carathéodory conditions on an open set $\Omega \subset \mathbb{R} \times C$ and a noncontinuable solution v of Eq. (2.2) on $[s-r, a)$, for each compact set $W \subset \Omega$ there exists $t_W \in [s - r, a)$ such that $(t, v_t) \notin W$ for $t \in [t_W, a)$.*

Proof. Otherwise, it would exist a compact set $W \subset \Omega$ and a sequence of real numbers $t_m \nearrow a$ when $m \to \infty$ such that $(t_m, v_{t_m}) \in W$ for all $m \in \mathbb{N}$. Since W is compact, it would also exist a subsequence t_{p_m} of t_m and $\varphi \in C$ such that

$$(t_{p_m}, v_{t_{p_m}}) \to (a, \varphi) \in W \quad \text{when} \quad m \to \infty.$$

In particular,

$$\lim_{m \to \infty} \sup_{\theta \in [-r,0]} |v_{t_{p_m}}(\theta) - \varphi(\theta)| = 0,$$

which implies that $v(a+\theta) = \varphi(\theta)$ for all $\theta \in [-r, 0)$. Hence, one can extend v to a continuous function on $[s - r, a]$ by letting $v(a) = \varphi(0)$. One can now apply Theorem 2.1 to find a solution w of Eq. (2.2) on some interval $[a - r, b)$ with $b > a$ and $w_a = \varphi$. But then the function $\bar{v}: [s - r, b) \to \mathbb{R}^n$ defined by

$$\bar{v}(t) = \begin{cases} v(t) & \text{if } t \in [s - r, a), \\ w(t) & \text{if } t \in [a, b) \end{cases}$$

is a solution of Eq. (2.2) that extends v to the interval $[s-r, b)$. In particular, v is continuable. This contradiction yields the desired statement. $\qquad\square$

2.2 Linear Equations and Evolution Families

In this section we consider a special class of nonautonomous linear delay equations. All delay equations considered in the book are either linear or are (seen as) perturbations of linear equations, which makes it natural to take the former as the starting point of our discussion. In particular, we introduce the notion of the evolution family associated with a linear delay equation and we discuss its bounded growth property.

Let $L(t): C \to \mathbb{R}^n$, for $t \in \mathbb{R}$, be bounded linear operators (sometimes it will be sufficient to consider linear operators $L(t)$ only for t sufficiently large, but the required changes are straightforward). Unless stated explicitly otherwise, we always assume that:

(1) the function $(t, v) \mapsto L(t)v$ satisfies the Carathéodory conditions on $\Omega = \mathbb{R} \times C$;
(2) the function

$$\Pi(t) = \|L(t)\| \tag{2.11}$$

is locally integrable, that is,

$$\int_s^t \Pi(\tau) \, d\tau < +\infty \quad \text{for } t > s.$$

We consider the *linear delay equation*

$$v' = L(t)v_t, \tag{2.12}$$

where v' is the right-hand derivative (when it exists). Given $(s, \varphi) \in \Omega$, a continuous function $v \colon [s - r, a) \to \mathbb{R}^n$ is a solution of the initial value problem

$$\begin{cases} v' = L(t)v_t, \\ v_s = \varphi \end{cases}$$

if and only if $v_s = \varphi$ and

$$v(t) = \varphi(0) + \int_s^t L(\tau)v_\tau \, d\tau \quad \text{for } t \in [s, a). \tag{2.13}$$

We observe that linear delay equations include linear ordinary differential equations as a particular case. Indeed, for $r = 0$ we have

$$L(t)\varphi = A(t)\varphi(0)$$

for some linear operators $A(t) \colon \mathbb{R}^n \to \mathbb{R}^n$. In this case Eq. (2.12) is the same as $v' = A(t)v$ since

$$L(t)v_t = A(t)v_t(0) = A(t)v(t).$$

Another particular class of linear delay equations is the class of linear differential-difference equations of the form

$$v' = A(t)v + \sum_{i=1}^k A_i(t)v(t - \tau_i(t)),$$

for some continuous functions τ_i with values in $[0, r]$, for $i = 1, \ldots, k$.

All linear delay equations have unique forward global solutions, with the standing assumptions in the beginning of the section.

Theorem 2.4 (Existence and uniqueness). *Given $(s, \varphi) \in \Omega$, there exists a unique solution $v \colon [s - r, +\infty) \to \mathbb{R}^n$ of Eq. (2.12) with $v_s = \varphi$.*

Proof. By Theorem 2.1, there exists a solution v of Eq. (2.12) on some interval $[s-r, a)$ with $v_s = \varphi$. We claim that v is continuable when $a < +\infty$. For $t \in [s, s + r]$, it follows from (2.13) that

$$\|v_t\| \le \max\left\{ \|v_s\|, \max_{\tau \in [s,t]} |v(\tau)| \right\}$$

$$\le \max\left\{ \|\varphi\|, |\varphi(0)| + \int_s^t \|L(\tau)v_\tau\| \, d\tau \right\} \tag{2.14}$$

$$\le \|\varphi\| + \int_s^t \|L(\tau)v_\tau\| \, d\tau.$$

On the other hand, for $t \in [s + r, a)$ we have

$$|v(t + \theta)| \leq |\varphi(0)| + \int_s^{t+\theta} \|L(\tau)v_\tau\| \, d\tau \tag{2.15}$$

for $\theta \in [-r, 0]$. Thus,

$$\|v_t\| \leq \|\varphi\| + \int_s^t \Pi(\tau)\|v_\tau\| \, d\tau \quad \text{for } t \in [s, a)$$

and it follows from Gronwall's lemma that

$$\|v_t\| \leq \|\varphi\| \exp\left(\int_s^t \Pi(\tau) \, d\tau\right) \quad \text{for } t \in [s, a). \tag{2.16}$$

Hence,

$$|v(\bar{t}) - v(t)| = \left|\int_t^{\bar{t}} L(\tau)v_\tau \, d\tau\right|$$

$$\leq \left|\int_t^{\bar{t}} \Pi(\tau) \, d\tau\right| \cdot \|\varphi\| \exp\left(\int_s^a \Pi(\tau) \, d\tau\right) \tag{2.17}$$

for any $t, \bar{t} \in [s, a)$ and so v is uniformly continuous on $[s - r, a)$ (since the function Π is locally integrable). Therefore, one can show as in the proof of Theorem 2.2 that

$$A = \{(t, v_t) : t \in [s, a)\}$$

is contained in a compact subset of Ω. It follows from Theorem 2.3 that the solution v is continuable (for any finite a) and so one can take $a = +\infty$. In other words, v can be continued to $[s - r, +\infty)$.

Now let w be another solution of Eq. (2.12) on the interval $[s - r, +\infty)$ with $w_s = \varphi$. Proceeding as in (2.14) and (2.15), we obtain

$$\|v_t - w_t\| \leq \int_s^t \Pi(\tau)\|v_\tau - w_\tau\| \, d\tau$$

for $t \geq s$. It follows from Gronwall's lemma that $v_t = w_t$ for $t \geq s$ and so $v = w$ on $[s - r, +\infty)$ (because $v_s = w_s = \varphi$). $\qquad\square$

Now we introduce the notion of an evolution family.

Definition 2.5. Given an interval $I \subset \mathbb{R}$, an *evolution family* on I is a family of bounded linear operators $T(t, s) \colon C \to C$, for $t, s \in I$ with $t \geq s$, such that

$$T(t, t) = \text{Id} \quad \text{and} \quad T(t, \tau)T(\tau, s) = T(t, s) \tag{2.18}$$

for $t, \tau, s \in I$ with $t \geq \tau \geq s$.

Using the solutions of Eq. (2.12), one can introduce a particular evolution family. Denote by

$$v_t = v_t(\cdot, s, \varphi), \quad \text{for } t \in [s, +\infty),$$

the unique solution given by Theorem 2.4.

Definition 2.6. Given an interval $I \subset \mathbb{R}$, the *evolution family* on I associated with Eq. (2.12) is the family of linear operators $T(t, s) \colon C \to C$, for $t, s \in I$ with $t \geq s$, defined by

$$T(t, s)\varphi = v_t(\cdot, s, \varphi) \quad \text{for } \varphi \in C. \tag{2.19}$$

By (2.16), each linear operator $T(t, s)$ in (2.19) is bounded. Moreover, the maps $T(t, s)$ satisfy the first identity in (2.18) and in view of the uniqueness of solutions, the second identity also holds: the solutions

$$t \mapsto T(t, \tau)T(\tau, s)\varphi \quad \text{and} \quad t \mapsto T(t, s)\varphi$$

of Eq. (2.12) coincide for $t = \tau$, which yields the desired property. In other words, any evolution family associated with a linear delay equation is indeed an evolution family.

By (2.16) we have the following bound for the norm of the evolution family.

Proposition 2.1. *The evolution family $T(t, s)$ associated with Eq. (2.12) satisfies*

$$\|T(t, s)\| \leq \exp\left(\int_s^t \Pi(\tau)\, d\tau\right) \quad \text{for } t \geq s.$$

One can use this bound to show that the solutions depend continuously on the initial data.

Proposition 2.2. *Let $T(t, s)$ be the evolution family associated with Eq. (2.12). Then the map $(t, s, \varphi) \mapsto T(t, s)\varphi$ is continuous on the set*

$$\{(t, s) \in \mathbb{R} \times \mathbb{R} : t \geq s\} \times C.$$

Proof. Take $\varphi, \psi \in C$ and $t, \bar{t}, s, \bar{s} \in \mathbb{R}$ with $t \geq s$ and $\bar{t} \geq \bar{s}$. Then

$$\begin{aligned}
T(t, s)\varphi - T(\bar{t}, \bar{s})\psi &= T(t, s)(\varphi - \psi) + T(t, \bar{t})T(\bar{t}, s)\psi - T(\bar{t}, s)T(s, \bar{s})\psi \\
&= T(t, s)(\varphi - \psi) + (T(t, \bar{t}) - \mathrm{Id})T(\bar{t}, s)\psi \\
&\quad + T(\bar{t}, s)(\mathrm{Id} - T(s, \bar{s}))\psi.
\end{aligned} \tag{2.20}$$

Now we observe that

$$\big(T(s,\bar{s})\psi\big)(\theta) = \psi(s - \bar{s} + \theta)$$

for $s - \bar{s} + \theta \leq 0$ and

$$\big(T(s,\bar{s})\psi\big)(\theta) = \psi(0) + \int_{\bar{s}}^{s+\theta} L(\tau)T(\tau,\bar{s})\psi \, d\tau$$

for $s - \bar{s} + \theta > 0$. Hence, by Proposition 2.1,

$$\|T(s,\bar{s})\psi - \psi\| \leq \sup_{\theta \in [-r,\bar{s}-s]} |\psi(s - \bar{s} + \theta) - \psi(\theta)|$$
$$+ \int_{\bar{s}}^{s} \Pi(\tau) \exp\left(\int_{\bar{s}}^{\tau} \Pi(\sigma) \, d\sigma\right) \|\psi\| \, d\tau$$
$$\leq \sup_{\theta \in [-r,\bar{s}-s]} |\psi(s - \bar{s} + \theta) - \psi(\theta)|$$
$$+ \int_{\bar{s}}^{s} \Pi(\tau) \, d\tau \exp\left(\int_{\bar{s}}^{s} \Pi(\sigma) \, d\sigma\right) \|\psi\|.$$

Finally, using (2.20), we obtain

$$\|T(t,s)\varphi - T(\bar{t},\bar{s})\psi\| \leq \exp\left(\int_{s}^{t} \Pi(\tau) \, d\tau\right) \|\varphi - \psi\|$$
$$+ \sup_{\theta \in [-r,\bar{t}-t]} |(T(\bar{t},s)\psi)(t - \bar{t} + \theta) - (T(\bar{t},s)\psi)(\theta)|$$
$$+ \int_{\bar{t}}^{t} \Pi(\tau) \, d\tau \exp\left(\int_{\bar{t}}^{t} \Pi(\tau) \, d\tau\right) \|T(\bar{t},s)\psi\|$$
$$+ \|T(\bar{t},s)\| \left[\sup_{\theta \in [-r,\bar{s}-s]} |\psi(s - \bar{s} + \theta) - \psi(\theta)| \right.$$
$$\left. + \int_{\bar{s}}^{s} \Pi(\tau) \, d\tau \exp\left(\int_{\bar{s}}^{s} \Pi(\sigma) \, d\sigma\right) \|\psi\| \right].$$

It follows readily from this inequality that if $(t,s) \to (\bar{t},\bar{s})$ and $\varphi \to \psi$, then

$$T(t,s)\varphi \to T(\bar{t},\bar{s})\psi \quad \text{in } C.$$

This yields the desired continuity property. $\square$

For the following theorem we recall that a linear operator is said to be *compact* if it takes bounded sets into relatively compact sets.

Theorem 2.5 (Compactness). *Assume that $r > 0$. For the evolution family on an interval I associated with Eq. (2.12), the operators $T(t,s)$ are compact for all $t, s \in I$ with $t \geq s + r$.*

Proof. Consider the ball

$$B = \{\varphi \in C : \|\varphi\| \le \rho\}$$

and take $\varphi \in B$ and $t, s \in I$ with $t \ge s + r$. By Proposition 2.1, we have

$$\|T(t,s)\varphi\| \le \exp\left(\int_s^t \Pi(\tau)\,d\tau\right)\|\varphi\| \le \exp\left(\int_s^t \Pi(\tau)\,d\tau\right)\rho =: \rho'.$$

On the other hand, by (2.17), for $t \ge s + r$ the function $\psi = T(t,s)\varphi$ satisfies

$$
\begin{aligned}
|\psi(\theta) - \psi(\bar{\theta})| &\le \left|\int_{t+\theta}^{t+\bar{\theta}} \Pi(\tau)\,d\tau\right| \cdot \|\varphi\| \exp\left(\int_s^t \Pi(\tau)\,d\tau\right) \\
&\le \rho|G(\theta) - G(\bar{\theta})|
\end{aligned}
$$

for all $\theta, \bar{\theta} \in [-r, 0]$, where

$$G(\theta) = \int_s^{t+\theta} \Pi(\tau)\,d\tau \exp\left(\int_s^t \Pi(\tau)\,d\tau\right).$$

Therefore, $T(t,s)B \subset B'$, where

$$B' = \left\{\psi \in C : \|\psi\| \le \rho', |\psi(\theta) - \psi(\bar{\theta})| \le \rho|G(\theta) - G(\bar{\theta})| \text{ for } \theta, \bar{\theta} \in [-r, 0]\right\}.$$

Since $r > 0$, it follows from the Arzelà–Ascoli theorem that B' is (sequentially) compact. This shows that the linear operator $T(t,s)$ is compact. $\square$

We end this section with a brief discussion of the bounded growth property.

Definition 2.7. An evolution family $T(t,s)$ on an interval $I \subset \mathbb{R}$ is said to have *bounded growth on I* if there exist constants $\omega, K > 0$ such that

$$\|T(t,s)\| \le Ke^{\omega(t-s)} \tag{2.21}$$

for $t, s \in I$ with $t \ge s$.

Definition 2.8. We say that Eq. (2.12) has *bounded growth on I* if its evolution family has bounded growth on this interval.

The next result is an easy consequence of Proposition 2.1.

Proposition 2.3. *If*

$$\sup_{t\in\mathbb{R}} \int_t^{t+1} \Pi(\tau)\,d\tau < +\infty, \tag{2.22}$$

then Eq. (2.12) has bounded growth on any interval.

2.3   Extension to Discontinuous Functions

In fact one can extend the domain of the evolution family $T(t,s)$ associated with Eq. (2.12) to certain discontinuous functions and this turns out to be a crucial development for the formulation of the variation of constants formula in Sec. 2.4 and for some applications in connection with the notion of hyperbolicity introduced in Chap. 3.

Let $\mathcal{L}(\mathbb{R}^n)$ be the set of all $n \times n$ matrices. We observe that the linear operator $L(t)\colon C \to \mathbb{R}^n$ can be written as the Riemann–Stieltjes integral

$$L(t)\varphi = \int_{-r}^{0} d\eta(t,\theta)\varphi(\theta) = \int_{-r}^{0} d_{\theta}\eta(t,\theta)\varphi(\theta) \qquad (2.23)$$

for some measurable map

$$\eta \colon \mathbb{R} \times [-r,0] \to \mathcal{L}(\mathbb{R}^n)$$

such that $\theta \mapsto \eta(t,\theta)$ has bounded variation and is left-continuous for each $t \in \mathbb{R}$. The function Π in (2.11) satisfies

$$\Pi(t) = \operatorname{Var}\eta(t,\cdot) \quad \text{for } t \in \mathbb{R},$$

where Var is the total variation on $[-r,0]$. For any bounded function $\varphi \colon [-r,0] \to \mathbb{R}^n$ that is Riemann–Stieltjes integrable with respect to $\eta(t,\cdot)$ we have

$$\left| \int_{-r}^{0} d\eta(t,\theta)\varphi(\theta) \right| \le \Pi(t)\|\varphi\|. \qquad (2.24)$$

Given $\alpha \in [-r,0]$, let $C_{\alpha} \supset C$ be the set of all functions $\varphi \colon [-r,0] \to \mathbb{R}^n$ with at most one point of discontinuity at α such that the limits

$$\lim_{\theta \to \alpha^-} \varphi(\theta) \quad \text{and} \quad \lim_{\theta \to \alpha^+} \varphi(\theta) = \varphi(\alpha)$$

exist (in particular φ is right-continuous at α). Note that all functions in C_{α} are bounded. In particular, $C_{-r} = C$ and C_0 is the set of all functions $\varphi \colon [-r,0] \to \mathbb{R}^n$ that are continuous on $[-r,0)$ for which the limit

$$\varphi(0^-) = \lim_{\theta \to 0^-} \varphi(\theta)$$

exists. Each set C_{α} endowed with the supremum norm in (2.1) is again a Banach space. Since each function $\varphi \in C_{\alpha}$ is right-continuous, it is Riemann–Stieltjes integrable with respect to $\eta(t,\cdot)$ for each $t \in \mathbb{R}$ (since the latter is left-continuous). Hence, one can extend each linear operator $L(t)$ to C_{α} using the integral in (2.23) and since there is no danger of confusion we continue to denote the extension by $L(t)$. We note that

$$\|L(t)|C\| = \|L(t)|C_{\alpha}\| = \Pi(t) \qquad (2.25)$$

for any $\alpha \in [-r, 0]$. This follows from the fact that $\|L(t)|C\| = \Pi(t)$ (the norm of a bounded linear operator on C is the total variation of any function of bounded variation that represents the linear operator as in (2.23)) together with inequality (2.24) for $\varphi \in C_\alpha$.

We also extend the notion of a solution in Definition 2.2 to certain discontinuous functions. On purpose, we restrict ourselves to linear equations.

Definition 2.9. A function $v \colon [s - r, a) \to \mathbb{R}^n$ with $a \le +\infty$ is called a *solution* of Eq. (2.12) if:

(1) v is continuous on $[s - r, a) \setminus \{s\}$ and the limits

$$\lim_{t \to s^-} v(t) \quad \text{and} \quad \lim_{t \to s^+} v(t) = v(s)$$

exist;

(2) we have

$$v(t) = v(s) + \int_s^t L(\tau)v_\tau \, d\tau \quad \text{for } t \in [s, a). \tag{2.26}$$

Property (2.26) can be written in the form

$$v(t) = v(s) + \int_s^t \int_{-r}^0 d_\theta \eta(\tau, \theta)v(\tau + \theta) \, d\tau \quad \text{for } t \in [s, a).$$

We note that any *continuous* solution of Eq. (2.12), in the sense of Definition 2.2, is also a solution in the sense of the more general Definition 2.9.

All linear delay equations have unique forward global solutions, with the standing assumptions in the beginning of Sec. 2.2.

Theorem 2.6 (Existence and uniqueness). *Given $(s, \varphi) \in \mathbb{R} \times C_0$, there exists a unique solution $v \colon [s-r, +\infty) \to \mathbb{R}^n$ of Eq. (2.12) with $v_s = \varphi$.*

Proof. Given $\varphi \in C_0$, define a continuous function $\psi \colon [s-r, +\infty) \to \mathbb{R}^n$ by

$$\psi_s = \varphi, \quad \psi(t) = \varphi(0) \text{ for } t \ge s.$$

Then v is a solution of Eq. (2.12) with $v_s = \varphi$ if and only if $u = v - \psi$ is a (continuous) solution (in the sense of Definition 2.2) of the initial value problem

$$\begin{cases} u' = L(t)u_t + g(t), \\ u_s = 0, \end{cases} \tag{2.27}$$

where

$$g(t) = L(t)\psi_t = \int_{-r}^0 d\eta(t, \theta)\psi(t + \theta).$$

Note that the function

$$\mathbb{R} \times C \ni (t, u) \mapsto L(t)u + g(t) \in \mathbb{R}^n$$

satisfies the Carathéodory conditions. The measurability in the first condition holds automatically, while the continuity follows from noting that

$$|L(t)u - L(t)\overline{u}| \le \Pi(t)\|u - \overline{u}\|$$

for all $t \in \mathbb{R}$ and $u, \overline{u} \in C$. Moreover,

$$|L(t)u + g(t)| \le \Pi(t)(\|u\| + \|\varphi\|),$$

which implies the third condition in Definition 2.1. Hence, by Theorem 2.1 there exists a solution u of problem (2.27) on some interval $[s - r, a)$.

We claim that u can be continued to the interval $[s - r, +\infty)$ and that it is the unique solution. For $t \in [s, s + r]$, since $u_s = 0$ we have

$$u(t) = \int_s^t L(\tau)u_\tau + \int_s^t g(\tau)\, d\tau$$

and so

$$\|u_t\| = \max_{\tau \in [s,t]} |u(\tau)| \le \int_s^t \Pi(\tau)\|u_\tau\|\, d\tau + \int_s^t |g(\tau)|\, d\tau.$$

On the other hand, for $t \in [s + r, a)$ we have

$$u(t + \theta) \le \int_s^{t+\theta} \Pi(\tau)\|u_\tau\|\, d\tau + \int_s^{t+\theta} |g(\tau)|\, d\tau$$

for $\theta \in [-r, 0]$. Thus,

$$\|u_t\| \le \int_s^a |g(\tau)|\, d\tau + \int_s^t \Pi(\tau)\|u_\tau\|\, d\tau.$$

By Gronwall's lemma we obtain

$$\|u_t\| \le \int_s^a |g(\tau)|\, d\tau \exp\left(\int_s^t \Pi(\tau)\, d\tau\right)$$

for $t \in [s, a)$. Therefore,

$$|L(t)u_t + g(t)| \le \Pi(t)\|u_t\| + |g(t)|$$

$$\le \Pi(t) \int_s^a |g(\tau)|\, d\tau \exp\left(\int_s^a \Pi(\tau)\, d\tau\right) + |g(t)|$$

for $t \in [s, a)$. Since Π and g are locally integrable, it follows as in (2.17) that u is uniformly continuous. One can now proceed as in the proof of Theorem 2.2 to show that

$$A = \{(t, u_t) : t \in [s, a)\}$$

is contained in a compact set. Hence, it follows from Theorem 2.3 that u is continuable (for any finite a) and so one can take $a = +\infty$.

For the uniqueness, we note that if $\overline{u}$ is another solution of problem (2.27) on the interval $[s - r, +\infty)$ with $\overline{u}_s = 0$, then

$$u(t) - \overline{u}(t) = \int_s^t L(\tau)(u_\tau - \overline{u}_\tau) \, d\tau$$

and so

$$\|u_t - \overline{u}_t\| \leq \int_s^t \Pi(\tau) \|u_\tau - \overline{u}_\tau\| \, d\tau.$$

Using once more Gronwall's lemma we conclude that $u = \overline{u}$ on $[s, +\infty)$ and so also on $[s - r, +\infty)$ since the initial conditions of u and $\overline{u}$ are the same.

Therefore, $v = u + \psi$ is the unique solution of Eq. (2.12) with $v_s = \varphi$, which concludes the proof of the theorem.   □

We continue to denote by $v_t = v_t(\cdot, s, \varphi)$ the unique solution of Eq. (2.12) with $v_s = \varphi \in C_0$.

Definition 2.10. For each $t \geq s$, we define a linear operator $T_0(t, s) \colon C_0 \to C_{\max\{s-t,-r\}}$ by

$$T_0(t, s)\varphi = v_t(\cdot, s, \varphi) \quad \text{for } \varphi \in C_0.$$

It follows readily from the definitions that

$$T_0(t, s)|C = T(t, s)$$

and

$$T_0(t, s)C_0 \subset C \quad \text{for } t \geq s + r. \tag{2.28}$$

The last property follows from observing that

$$(T_0(t, s)v_s)(\theta) = v(t + \theta),$$

since $t + \theta \geq s$ for $t \geq s + r$ (recall that v has at most one point of discontinuity at s).

Note that $T_0(t, s)$ is not an evolution family, since the spaces C_0 and $C_{\max\{s-t,-r\}}$ in Definition 2.10 need not be the same. However, one can extend the existence and uniqueness of solutions in Theorem 2.6 to the initial conditions (s, φ) in the set $\mathbb{R} \times C_\alpha$. This allows one to define linear operators

$$T_\alpha(t, s) \colon C_\alpha \to C_{\max\{s-t+\alpha,-r\}} \quad \text{for } t \geq s$$

by

$$T_\alpha(t,s)\varphi = v_t(\cdot, s, \varphi) \quad \text{for } \varphi \in C_\alpha.$$

Then

$$T_\beta(t,\tau)T_\alpha(\tau,s) = T_\alpha(t,s) \quad \text{for } t \geq \tau \geq s,$$

where $\beta = \max\{s-\tau+\alpha, -r\}$, with the convention that $T_{-r}(t,\tau) = T(t,\tau)$.

Finally, we obtain a bounded growth property for the norms

$$\|T_0(t,s)\| = \sup_{\varphi \in C_0 \backslash \{0\}} \frac{\|T_0(t,s)\varphi\|}{\|\varphi\|}$$

of the linear operators $T_0(t,s)$.

Proposition 2.4. *If condition* (2.22) *holds, then there exist constants* $\omega, K > 0$ *such that*

$$\|T_0(t,s)\| \leq Ke^{\omega(t-s)} \quad \text{for } t \geq s. \tag{2.29}$$

Proof. Repeating the proof of Theorem 2.4 taking $\varphi \in C_0$ and $a = +\infty$, it follows from (2.25) together with (2.16) that

$$\|v_t\| \leq \|\varphi\| \exp\left(\int_s^t \Pi(\tau)\, d\tau\right) \quad \text{for } t \geq s, \tag{2.30}$$

which is equivalent to

$$\|T_0(t,s)\| \leq \exp\left(\int_s^t \Pi(\tau)\, d\tau\right) \quad \text{for } t \geq s.$$

The desired property follows now immediately from (2.22). $\qquad\square$

We note that the constants ω and K in (2.29) are indeed the same as those for the evolution family $T(t,s)$ in Proposition 2.3.

2.4 Variation of Constants Formula

In this section we establish the variation of constants formula for the perturbations of a nonautonomous linear delay equation. More precisely, we consider the perturbed equation

$$v' = L(t)v_t + g(t), \tag{2.31}$$

with the standing assumptions in the beginning of Sec. 2.2 and where $g\colon \mathbb{R} \to \mathbb{R}^n$ is a locally integrable function.

We first discuss the existence and uniqueness of solutions.

Proposition 2.5 (Existence and uniqueness). *Given $(s, \varphi) \in \Omega$, there exists a unique solution $v \colon [s - r, +\infty) \to \mathbb{R}^n$ of Eq. (2.31) with $v_s = \varphi$.*

Proof. First observe that the function
$$f(t, v) = L(t)v + g(t)$$
satisfies the Carathéodory conditions (see Definition 2.1). In particular, the third condition follows readily from noting that
$$|f(t, v)| \le \Pi(t)\|v\| + g(t).$$
Hence, by Theorem 2.1, there exists a solution v of Eq. (2.31) on $[s - r, a)$.

Proceeding as in (2.14) and (2.15) with $L(\tau)v_\tau$ replaced by $f(\tau, v_\tau)$, we obtain
$$\|v_t\| \le \|\varphi\| + \int_s^t |g(\tau)|\, d\tau + \int_s^t \Pi(\tau)\|v_\tau\|\, d\tau$$
for $t \in [s, a)$. Applying Gronwall's lemma we conclude that
$$\|v_t\| \le \left(\|\varphi\| + \int_s^a |g(\tau)|\, d\tau\right) \exp\left(\int_s^t \Pi(\tau)\, d\tau\right)$$
for $t \in [s, a)$ and so also
$$|L(t)v_t + g(t)| \le \Pi(t)\left(\|\varphi\| + \int_s^a |g(\tau)|\, d\tau\right) \exp\left(\int_s^a \Pi(\tau)\, d\tau\right) + |g(t)|$$
for $t \in [s, a)$. Since Π and g are locally integrable, it follows as in (2.17) that v is uniformly continuous. One can also show as in the proof of Theorem 2.2 that
$$A = \{(t, v_t) : t \in [s, a)\}$$
is contained in a compact set and, by Theorem 2.3, v can be continued to the interval $[s - r, +\infty)$.

For the uniqueness, note that if w is another solution of Eq. (2.31) on the interval $[s - r, +\infty)$ with $w_s = \varphi$, then
$$\|v_t - w_t\| \le \int_s^t \Pi(\tau)\|v_\tau - w_\tau\|\, d\tau \quad \text{for } t \ge s.$$
Hence, by Gronwall's lemma, we conclude that $v = w$ on $[s, +\infty)$ and so also on $[s - r, +\infty)$. $\qquad\square$

Now we are ready to establish the variation of constants formula. Define a linear operator $X_0 \colon \mathbb{R}^n \to C_0$ by
$$(X_0 p)(\theta) = \begin{cases} 0 & \text{if } -r \le \theta < 0, \\ p & \text{if } \theta = 0 \end{cases} \tag{2.32}$$
for each $p \in \mathbb{R}^n$.

Theorem 2.7 (Variation of constants formula). *Given* $(s, \varphi) \in \Omega$, *a function* $v \colon [s - r, +\infty) \to \mathbb{R}^n$ *is the solution of Eq. (2.31) with* $v_s = \varphi$ *if and only if*

$$v_t = T(t, s)\varphi + \int_s^t T_0(t, \tau) X_0 g(\tau) \, d\tau \quad \text{for } t \geq s, \qquad (2.33)$$

in the sense that $v_s = \varphi$ *and*

$$v(t + \theta) = (T(t, s)\varphi)(\theta) + \int_s^{t+\theta} \big(T_0(t, \tau) X_0 g(\tau)\big)(\theta) \, d\tau \qquad (2.34)$$

for $t \geq s$ *and* $\theta \in [-r, 0]$ *with* $t + \theta \geq s$.

Proof. We define matrices $R(t, \tau) \in \mathcal{L}(\mathbb{R}^n)$ for $t, \tau \geq s$ by

$$R(t, \tau)p = \begin{cases} (T_0(t, \tau) X_0 p)(0) & \text{if } t \geq \tau, \\ 0 & \text{if } t < \tau \end{cases} \qquad (2.35)$$

for each $p \in \mathbb{R}^n$. The following result is a first characterization of the solutions of Eq. (2.31).

Lemma 2.1. *Given* $(s, \varphi) \in \Omega$, *the solution of Eq. (2.31) with* $v_s = \varphi$ *is given by*

$$v(t) = (T(t, s)\varphi)(0) + \int_s^t R(t, \tau) g(\tau) \, d\tau \quad \text{for } t \geq s. \qquad (2.36)$$

Proof of the lemma. Consider the function $u \colon [s - r, +\infty) \to \mathbb{R}^n$ such that $u_s = 0$ and

$$u(t) = \int_s^t R(t, \tau) g(\tau) \, d\tau \quad \text{for } t \geq s. \qquad (2.37)$$

By (2.30) we have

$$\int_s^t |R(t, \tau) g(\tau)| \, d\tau \leq \int_s^t |g(\tau)| \exp\left(\int_\tau^t \Pi(\sigma) \, d\sigma\right) d\tau$$

$$\leq \int_s^t |g(\tau)| \, d\tau \exp\left(\int_s^t \Pi(\sigma) \, d\sigma\right) < +\infty$$

and so the integral in (2.37) is well defined. For each $t \geq \tau$, since

$$R(t, \tau)p = p + \int_\tau^t L(\sigma)(T_0(\sigma, \tau) X_0 p) \, d\sigma$$

$$= p + \int_\tau^t \int_{-r}^0 d\eta(\sigma, \theta)(T_0(\sigma, \tau) X_0 p)(\theta) \, d\sigma$$

$$= p + \int_\tau^t \int_{-r}^0 d\eta(\sigma, \theta) R(\sigma + \theta, \tau) p \, d\sigma,$$

we have

$$R(t, \tau) = \mathrm{Id} + \int_\tau^t \int_{-r}^0 d\eta(\sigma, \theta) R(\sigma + \theta, \tau) \, d\sigma.$$

Hence,

$$u(t) = \int_s^t g(\tau) \, d\tau + \int_s^t \int_\sigma^t \int_{-r}^0 d\eta(\tau, \theta) R(\tau + \theta, \sigma) \, d\tau \, g(\sigma) \, d\sigma$$

and so

$$
\begin{aligned}
u(t) - \int_s^t g(\tau) \, d\tau &= \int_s^t \int_\sigma^t \int_{-r}^0 d\eta(\tau, \theta) R(\tau + \theta, \sigma) \, d\tau \, g(\sigma) \, d\sigma \\
&= \int_s^t \int_{-r}^0 d\eta(\tau, \theta) \int_s^\tau R(\tau + \theta, \sigma) g(\sigma) \, d\sigma \, d\tau \\
&= \int_s^t \int_{-r}^0 d\eta(\tau, \theta) \int_s^{\tau + \theta} R(\tau + \theta, \sigma) g(\sigma) \, d\sigma \, d\tau \\
&= \int_s^t \int_{-r}^0 d\eta(\tau, \theta) u(\tau + \theta) \, d\tau \\
&= \int_s^t L(\tau) u_\tau \, d\tau.
\end{aligned}
$$

This shows that

$$u(t) = \int_s^t \big(L(\tau) u_\tau + g(\tau) \big) \, d\tau.$$

Now we consider the function $v \colon [s - r, +\infty) \to \mathbb{R}^n$ defined by

$$v(t) = (T(t, s)\varphi)(0) + u(t) \quad \text{for } t \geq s$$

such that $v_s = \varphi$. Since $t \mapsto T(t, s)\varphi$ is a solution of Eq. (2.12), we have

$$(T(t, s)\varphi)(0) - \varphi(0) = \int_s^t L(\tau) T(\tau, s)\varphi \, d\tau$$

and so

$$
\begin{aligned}
v(t) &= (T(t, s)\varphi)(0) + \int_s^t L(\tau) u_\tau \, d\tau + \int_s^t g(\tau) \, d\tau \\
&= \varphi(0) + \int_s^t L(\tau) \big(u_\tau + T(\tau, s)\varphi \big) \, d\tau + \int_s^t g(\tau) \, d\tau \\
&= \varphi(0) + \int_s^t L(\tau) v_\tau \, d\tau + \int_s^t g(\tau) \, d\tau
\end{aligned}
$$

for $t \geq s$. Hence, v is the solution of Eq. (2.31) with $v_s = \varphi$. $\qquad \square$

We proceed with the proof of the theorem. First let v be the solution of Eq. (2.31) with $v_s = \varphi$. Replacing t by $t + \theta$ in identity (2.36) with $\theta \in [-r, 0]$, we obtain

$$v(t + \theta) = (T(t + \theta, s)\varphi)(0) + \int_s^{t+\theta} R(t + \theta, \tau)g(\tau) \, d\tau$$

for $t + \theta \geq s$ and

$$v(t + \theta) = \varphi(t - s + \theta)$$

for $t + \theta \in [s - r, s]$. Moreover, by (2.35), we have

$$R(t + \theta, \tau)g(\tau) = (T_0(t + \theta, \tau)X_0 g(\tau))(0) = (T_0(t, \tau)X_0 g(\tau))(\theta).$$

On the other hand,

$$(T(t, s)\varphi)(\theta) = \begin{cases} (T(t + \theta, s)\varphi)(0) & \text{if } t + \theta \geq s, \\ \varphi(t - s + \theta) & \text{if } t + \theta \in [s - r, s]. \end{cases}$$

Hence,

$$v(t + \theta) = (T(t, s)\varphi)(\theta) + \int_s^{t+\theta} (T_0(t, \tau)X_0 g(\tau))(\theta) \, d\tau$$

for $t + \theta \geq s$ and so identity (2.33) holds, in the sense of (2.34).

Now assume that identity (2.34) holds. One can define a continuous function $v \colon [s - r, +\infty) \to \mathbb{R}^n$ by

$$v(t + \theta) = v_t(\theta) \quad \text{for } t \geq s, \theta \in [-r, 0]. \tag{2.38}$$

Indeed, note that

$$(T(t, s)\varphi)(\theta) = (T(t + \theta, s)\varphi)(0) \quad \text{for } t \geq s, \theta \in [-r, 0].$$

Moreover,

$$\int_s^{t+\theta} (T_0(t, \tau)X_0 g(\tau))(\theta) \, d\tau = \int_s^{t+\theta} (T_0(t + \theta, \tau)X_0 g(\tau))(0) \, d\tau$$

for $t + \theta \geq s$ and

$$\int_s^{t+\theta} (T_0(t, \tau)X_0 g(\tau))(\theta) \, d\tau = 0$$

for $t + \theta \leq s$. Therefore, the function $v \colon [s - r, +\infty) \to \mathbb{R}^n$ defined by

$$v(t) = \begin{cases} v_t(0) & \text{if } t \geq s, \\ \varphi(\theta) & \text{if } t = s + \theta, \theta \in [-r, 0] \end{cases}$$

satisfies property (2.38). Clearly, $v_s = \varphi$. It remains to show that v is a solution of Eq. (2.31). Taking $\theta = 0$ in (2.34), we obtain

$$v(t) = (T(t,s)\varphi)(0) + \int_s^t \big(T_0(t,\tau)X_0 g(\tau)\big)(0)\, d\tau$$

$$= (T(t,s)\varphi)(0) + \int_s^t R(t,\tau)g(\tau)\, d\tau$$

for $t \geq s$. Hence, it follows from Lemma 2.1 that v is the solution of Eq. (2.31) with $v_s = \varphi$. This completes the proof of the theorem.   □

For the solution $v \colon [s - r, +\infty) \to \mathbb{R}^n$ of Eq. (2.31) with $v_s = \varphi$, it follows readily from Theorem 2.7 that

$$v_t = T(t,\bar{s})v_{\bar{s}} + \int_{\bar{s}}^t T_0(t,\tau)X_0 g(\tau)\, d\tau \quad \text{for } t \geq \bar{s} \geq s,$$

in the sense that

$$v(t + \theta) = (T(t,\bar{s})v_{\bar{s}})(\theta) + \int_{\bar{s}}^{t+\theta} \big(T_0(t,\tau)X_0 g(\tau)\big)(\theta)\, d\tau \qquad (2.39)$$

for $t \geq \bar{s} \geq s$ and $\theta \in [-r, 0]$ with $t + \theta \geq \bar{s}$. Moreover, it follows from (2.39) that

$$|v(t + \theta)| \leq |(T(t,\bar{s})v_{\bar{s}})(\theta)| + \int_{\bar{s}}^{t+\theta} \big|\big(T_0(t,\tau)X_0 g(\tau)\big)(\theta)\big|\, d\tau$$

$$\leq \|T(t,\bar{s})v_{\bar{s}}\| + \int_{\bar{s}}^t \|T_0(t,\tau)\| \cdot |g(\tau)|\, d\tau$$

for $t + \theta \geq \bar{s}$ and so

$$\|v_t\| \leq \|T(t,\bar{s})v_{\bar{s}}\| + \int_{\bar{s}}^t \|T_0(t,\tau)\| \cdot |g(\tau)|\, d\tau \quad \text{for } t \geq \bar{s} \geq s.$$

2.5   Perturbations of Autonomous Equations

In this section we consider briefly the autonomous linear equation

$$v' = Lv_t, \qquad (2.40)$$

where $L \colon C \to \mathbb{R}^n$ is a bounded linear operator. We also consider a class of nonautonomous perturbations of the equation.

By Theorem 2.4, given $(s, \varphi) \in \mathbb{R} \times C$, there exists a unique solution

$$v_t = v_t(\cdot, s, \varphi), \quad \text{for } t \in [s, +\infty),$$

of Eq. (2.40) with $v_s = \varphi$. We define linear operators $S(t)$ for $t \geq 0$ by

$$S(t)\varphi = v_t(\cdot, 0, \varphi). \tag{2.41}$$

One can easily verify that the linear operators $T(t, s)$ in (2.19) satisfy

$$T(t, s) = S(t - s) \quad \text{for } t \geq s. \tag{2.42}$$

As a consequence, $S(t)$ is a semigroup, that is,

$$S(0) = \text{Id} \quad \text{and} \quad S(t)S(s) = S(t + s) \tag{2.43}$$

for all $t, s \geq 0$. Moreover, by Proposition 2.1 we have

$$\|S(t)\| \leq e^{\|L\|t} \quad \text{for } t \geq 0. \tag{2.44}$$

In a similar manner to that in Sec. 2.3, one can extend L to the space C_0 writing it as the Riemann–Stieltjes integral

$$L\varphi = \int_{-r}^{0} d\eta(\theta)\varphi(\theta), \tag{2.45}$$

where $\eta \colon \mathbb{R} \to \mathcal{L}(\mathbb{R}^n)$ has bounded variation and is left-continuous. This allows one to define linear operators $S_0(t)$ on C_0 by

$$S_0(t)\varphi = v_t(\cdot, 0, \varphi) \quad \text{for } \varphi \in C_0,$$

where $v_t(\cdot, s, \varphi)$ is the unique solution of the equation $v' = Lv_t$ with $v_s = \varphi$.

Finally, we present a variation of constants formula. It is an immediate consequence of Theorem 2.7 together with property (2.42).

Theorem 2.8 (Variation of constants formula). *Let $g \colon \mathbb{R} \to \mathbb{R}^n$ be a locally integrable function. Given $(s, \varphi) \in \Omega$, there exists a unique solution $v \colon [s - r, +\infty) \to \mathbb{R}^n$ of the equation $v' = Lv_t + g(t)$ with $v_s = \varphi$, given by*

$$v_t = S(t - s)\varphi + \int_{s}^{t} S_0(t - \tau)X_0 g(\tau)\, d\tau \quad \text{for } t \geq s,$$

in the sense that $v_s = \varphi$ and

$$v(t + \theta) = (S(t - s)\varphi)(\theta) + \int_{s}^{t+\theta} \big(S_0(t - \tau)X_0 g(\tau)\big)(\theta)\, d\tau$$

for $t \geq s$ and $\theta \in [-r, 0]$ with $t + \theta \geq s$.

For the solution v in Theorem 2.8, we also have

$$v_t = S(t - \bar{s})v_{\bar{s}} + \int_{\bar{s}}^{t} S_0(t - \tau)X_0 g(\tau)\, d\tau$$

and

$$\|v_t\| \leq \|S(t - \bar{s})v_{\bar{s}}\| + \int_{\bar{s}}^{t} \|S_0(t - \tau)\| \cdot |g(\tau)|\, d\tau$$

for all $t \geq \bar{s} \geq s$.

Chapter 3

Hyperbolicity

In this chapter we present the concept of hyperbolicity in the context of delay equations. We first introduce the notions of an exponential contraction and of an exponential dichotomy for a linear delay equation. In particular, we show that for equations with bounded growth any exponential contraction extends to certain discontinuous functions considered in Chap. 2 and that any exponential dichotomy can be characterized by the existence of contraction and expansion, without the need for information on the norms of the projections onto the stable and unstable spaces. We then present an equivalent description of the exponential behavior in terms of Lyapunov functions. We also show that for a linear equation any exponential dichotomy can be extended to certain discontinuous functions. As an application, we describe the projections of the variation of constants formula onto the stable and unstable spaces, which is crucial later on.

3.1 Exponential Behavior: Basic Notions

In this section we introduce the notions of an exponential contraction and of an exponential dichotomy for the linear delay Eq. (2.12), that is,

$$v' = L(t)v_t,$$

where $L(t) \colon C \to \mathbb{R}^n$, for $t \in \mathbb{R}$, are bounded linear operators satisfying the standing assumptions in the beginning of Sec. 2.2. Let $T(t,s)$ be the evolution family associated with Eq. (2.12) (see Definition 2.6).

Definition 3.1. Given an interval $I \subset \mathbb{R}$, we say that Eq. (2.12) has an *exponential contraction* on I if there exist constants $\lambda, D > 0$ such that

$$\|T(t,s)\| \leq De^{-\lambda(t-s)} \tag{3.1}$$

for $t, s \in I$ with $t \geq s$.

We show that if Eq. (2.12) has bounded growth, then any exponential contraction can be extended to the space C_0.

Proposition 3.1. *If Eq. (2.12) has bounded growth and has an exponential contraction on I, then there exists $D' > 0$ such that*

$$\|T_0(t, s)\| \leq D' e^{-\lambda(t-s)}$$

for $t, s \in I$ with $t \geq s$.

Proof. Given $\varphi \in C_0$, we have

$$\psi := T_0(s + r, s)\varphi \in C.$$

For $t - s \geq r$, it follows from Proposition 2.4 and (3.1) that

$$\begin{aligned}
\|T_0(t, s)\varphi\| &\leq D e^{-\lambda(t-s-r)}\|\psi\| \\
&\leq D e^{-\lambda(t-s-r)} K e^{\omega r}\|\varphi\| \\
&= DK e^{(\lambda+\omega)r} e^{-\lambda(t-s)}\|\varphi\|.
\end{aligned}$$

On the other hand, for $t - s \leq r$, we have $1 \leq e^{\lambda r - \lambda(t-s)}$ and it follows from Proposition 2.4 that

$$\begin{aligned}
\|T_0(t, s)\varphi\| &\leq K e^{\omega(t-s)}\|\varphi\| \leq K e^{\omega r}\|\varphi\| \\
&\leq K e^{(\lambda+\omega)r} e^{-\lambda(t-s)}\|\varphi\|.
\end{aligned}$$

This yields the desired result. $\qquad\square$

Now we consider the more general case when the dynamics has both contraction and expansion.

Definition 3.2. Given an interval $I \subset \mathbb{R}$, we say that Eq. (2.12) has an *exponential dichotomy* on I if:

(1) there exist projections $P(t): C \to C$ for $t \in I$ such that

$$P(t)T(t, s) = T(t, s)P(s) \quad \text{for } t \geq s; \tag{3.2}$$

(2) the linear operator

$$\overline{T}(t, s) = T(t, s)|\operatorname{Ker} P(s): \operatorname{Ker} P(s) \to \operatorname{Ker} P(t) \tag{3.3}$$

is invertible for $t \geq s$;

(3) there exist constants $\lambda, D > 0$ such that for every $t, s \in I$ with $t \geq s$ we have

$$\|T(t, s)P(s)\| \leq D e^{-\lambda(t-s)}, \quad \|\overline{T}(s, t)Q(t)\| \leq D e^{-\lambda(t-s)}, \tag{3.4}$$

where $\overline{T}(s, t) = \overline{T}(t, s)^{-1}$ and $Q(t) = \operatorname{Id} - P(t)$.

Note that Eq. (2.12) has an exponential contraction on I if and only if it has an exponential dichotomy on I with $P(t) = \mathrm{Id}$ for all $t \in I$.

Given an interval $I \subset \mathbb{R}$ and projections $P(t)$ for $t \in I$, we consider the spaces

$$E(t) = P(t)C \quad \text{and} \quad F(t) = Q(t)C,$$

where $Q(t) = \mathrm{Id} - P(t)$. Clearly,

$$C = E(t) \oplus F(t) \quad \text{for } t \in I.$$

Definition 3.3. When Eq. (2.12) has an exponential dichotomy on an interval I, the spaces $E(t)$ and $F(t)$ are called, respectively, *stable* and *unstable* spaces at $t \in I$.

The following result shows that when Eq. (2.12) has bounded growth, one can replace the third condition in the notion of an exponential dichotomy by the requirement that there exist contraction and expansion, respectively, along the spaces $E(t)$ and $F(t)$.

Theorem 3.1. *Assume that Eq. (2.12) has bounded growth and that conditions 1 and 2 in Definition 3.2 hold on an interval I containing $\mathbb{R}_0^+$. Then the equation has an exponential dichotomy on I if and only if there exist constants $\lambda, D' > 0$ such that*

$$\|T(t,s)|E(t)\| \le D'e^{-\lambda(t-s)}, \quad \|\overline{T}(s,t)\| \le D'e^{-\lambda(t-s)} \qquad (3.5)$$

for every $t, s \in I$ with $t \ge s$.

Proof. Assume that Eq. (2.12) has an exponential dichotomy. By (3.4) we have

$$\|T(t,s)\varphi\| \le De^{-\lambda(t-s)}\|\varphi\| \quad \text{and} \quad \|\overline{T}(s,t)\psi\| \le De^{-\lambda(t-s)}\|\psi\|$$

for $t \ge s$, $\varphi \in E(s)$ and $\psi \in F(t)$. This establishes the inequalities in (3.5).

Now assume that (3.5) holds and define

$$\Theta(s) = \inf\{\|x - y\| : x \in E(s), y \in F(s), \|x\| = \|y\| = 1\}. \qquad (3.6)$$

Take $\varphi \in C$ and let $x = P(s)\varphi$ and $y = Q(s)\varphi$. Then

$$\Theta(s) \le \left\| \frac{x}{\|x\|} - \frac{y}{\|y\|} \right\| = \frac{1}{\|x\|} \left\| x - \frac{\|x\|}{\|y\|}y \right\|$$

$$= \frac{1}{\|x\|} \left\| \varphi + \frac{\|P(s)\varphi\| - \|Q(s)\varphi\|}{\|y\|} y \right\| \le \frac{2\|\varphi\|}{\|P(s)\varphi\|}$$

and so
$$\|P(s)\| \leq \frac{2}{\Theta(s)}. \tag{3.7}$$

Now take $x \in E(s)$ and $y \in F(s)$ with $\|x\| = \|y\| = 1$, as in (3.6). By the bounded growth property we have
$$\|T(t,s)(x-y)\| \leq Ke^{\omega(t-s)}\|y-x\|.$$

Therefore,
$$\|y-x\| \geq K^{-1}e^{-\omega(t-s)}\left(D^{-1}e^{\omega(t-s)} - De^{-\lambda(t-s)}\right) \tag{3.8}$$

for all $t \geq s$. Taking $\tau > 0$ such that
$$\rho := K^{-1}e^{-\omega\tau}(D^{-1}e^{\lambda\tau} - De^{-\lambda\tau}) > 0$$

and $t = s + \tau$ (which is in I since $I \supset \mathbb{R}_0^+$), it follows from (3.8) that
$$\Theta(s) \geq \rho \tag{3.9}$$

for $s \in I$. Hence, by (3.7), we have $\|P(s)\| \leq 2/\rho$ for $s \in I$. A similar bound can be obtained for $Q(t)$ or, alternatively, one can note that
$$\|Q(s)\| = \|\mathrm{Id} - P(s)\| \leq 1 + \|P(s)\|,$$

again for all $s \in I$. Since
$$\|T(t,s)P(s)\| \leq \|T(t,s)|E(s)\| \cdot \|P(s)\|$$

and
$$\|\overline{T}(s,t)Q(t)\| \leq \|\overline{T}(s,t)\| \cdot \|Q(t)\|,$$

we conclude that Eq. (2.12) has an exponential dichotomy. $\square$

Finally, we show that the unstable spaces of an exponential dichotomy are finite-dimensional whenever $r > 0$.

Theorem 3.2. *Assume that $r > 0$. If Eq. (2.12) has an exponential dichotomy on an interval I containing $\mathbb{R}_0^+$, then $\dim F(t) < \infty$ for all $t \in I$.*

Proof. Take $t \geq r$ with $e^{\lambda t} > D$ and let
$$B(t) = \{\varphi \in F(t) : \|\varphi\| \leq 1\}.$$

Given $\varphi \in B(t)$, there exists $\psi \in F(0)$ such that $\varphi = T(t,0)\psi$. Assume that $\|\psi\| > 1$. It follows from (3.4) that
$$1 < \frac{1}{D}e^{\lambda t}\|\psi\| \leq \|T(t,0)\psi\| = \|\varphi\|,$$

although $\varphi \in B(t)$. This contradiction shows that $\|\psi\| \leq 1$ and so
$$B(t) \subset T(t,0)B(0).$$

Since $B(0)$ is bounded and $T(t,0)$ is compact (by Theorem 2.5), the set $T(t,0)B(0)$ is relatively compact and thus, $B(t)$ is compact. This implies that $F(t)$ is finite-dimensional for any sufficiently large t. On the other hand, it follows from the invertibility of the map $\overline{T}(t,0)$ that $\dim F(t)$ is the same for all $t \in I$. $\square$

3.2   Characterization via Lyapunov Functions

In this section we show how the existence of an exponential contraction or an exponential dichotomy can be characterized in terms of strict Lyapunov functions. This provides an alternative approach to verify the existence of exponential behavior.

We first consider exponential contractions. We continue to denote by $T(t, s)$ the evolution family associated with Eq. (2.12) (see Definition 2.6).

Definition 3.4. Given an interval $I \subset \mathbb{R}$, a function $V \colon I \times C \to \mathbb{R}_0^-$ is called a *strict Lyapunov function* for Eq. (2.12) on I if there exist constants $\gamma, d > 0$ such that

$$\gamma^{-1}\|\varphi\| \le |V(t, \varphi)| \le \gamma\|\varphi\| \tag{3.10}$$

and

$$|V(t, T(t, s)\varphi)| \le e^{-d(t-s)}|V(s, \varphi)| \tag{3.11}$$

for every $t, s \in I$ with $t \ge s$ and $\varphi \in C$.

The following result gives an optimal characterization of an exponential contraction in terms of the existence of a Lyapunov function.

Proposition 3.2. *The following properties are equivalent:*

(1) Eq. (2.12) has an exponential contraction on I;
(2) there exists a strict Lyapunov function for Eq. (2.12) on I.

Proof. We first assume that V is a strict Lyapunov function for Eq. (2.12). By conditions (3.10) and (3.11), for every $t \ge s$ and $\varphi \in C$ we have

$$\begin{aligned}
\|T(t, s)\varphi\| &\le \gamma|V(t, T(t, s)\varphi)| \\
&\le \gamma e^{-d(t-s)}|V(s, \varphi)| \\
&\le \gamma^2 e^{-d(t-s)}\|\varphi\|
\end{aligned}$$

and so Eq. (2.12) has an exponential contraction.

Now assume that the equation has an exponential contraction. We construct explicitly a strict Lyapunov function. For each $t \in I$ and $\varphi \in C$, let

$$V(t, \varphi) = -\sup\{\|T(\tau, t)\varphi\|e^{\lambda(\tau - t)} : \tau \ge t\}.$$

It follows from (3.1) that

$$|V(t, \varphi)| = \sup\{\|T(\tau, t)\varphi\|e^{\lambda(\tau - t)} : \tau \ge t\} \le D\|\varphi\|$$

and taking $t = \tau$ we obtain $|V(t, \varphi)| \geq \|\varphi\|$. This establishes condition (3.10). Moreover, for each $t \geq s$ we have

$$
\begin{aligned}
|V(t, T(t, s)\varphi)| &= \sup\{\|T(\tau, t)T(t, s)\varphi\|e^{\lambda(\tau - t)} : \tau \geq t\} \\
&= e^{-\lambda(t-s)} \sup\{\|T(\tau, s)\varphi\|e^{\lambda(\tau - s)} : \tau \geq t\} \\
&\leq e^{-\lambda(t-s)} \sup\{\|T(\tau, s)\varphi\|e^{\lambda(\tau - s)} : \tau \geq s\} \\
&= e^{-\lambda(t-s)}|V(s, \varphi)|
\end{aligned}
$$

and so condition (3.11) also holds. $\qquad\square$

For a differentiable strict Lyapunov function, condition (3.11) can be reformulated as follows. Given a differentiable function $V \colon I \times C \to \mathbb{R}_0^-$, let

$$
\dot{V}(t, \varphi) = \lim_{h \searrow 0} \frac{V(t + h, T(t + h, t)\varphi) - V(t, \varphi)}{h}.
$$

Then condition (3.11) holds if and only if

$$
\dot{V}(t, \varphi) \geq -dV(t, \varphi) \quad \text{for } t \in I, \varphi \in C.
$$

Now we consider exponential dichotomies and we obtain a corresponding description in terms of strict Lyapunov functions.

Theorem 3.3. *Assume that Eq. (2.12) has bounded growth and that conditions 1 and 2 in Definition 3.2 hold on an interval I containing $\mathbb{R}_0^+$. Then the equation has an exponential dichotomy on I if and only if there exists a function $V \colon I \times C \to \mathbb{R}$ such that:*

(1) $V(t, \varphi) \leq 0$ for $\varphi \in E(t)$ and $V(t, \varphi) > 0$ for $\varphi \in F(t)$;
(2) there exist constants $\gamma, d > 0$ such that

$$
\gamma^{-1}\|\varphi\| \leq |V(t, \varphi)| \leq \gamma\|\varphi\| \quad \text{for } \varphi \in E(t) \cup F(t) \tag{3.12}
$$

and

$$
|V(t, T(t, s)\varphi)| \leq e^{-d(t-s)}|V(s, \varphi)|, \quad V(s, \overline{T}(s, t)\psi) \leq e^{-d(t-s)}V(t, \psi) \tag{3.13}
$$

for $t \geq s$, $\varphi \in E(s)$ and $\psi \in F(t)$.

Proof. We first assume that there exists a function V satisfying conditions 1 and 2. Using (3.12) and (3.13), one can repeat the proof of Proposition 3.2 to conclude that condition (3.5) holds. Hence, it follows from Theorem 3.1 that Eq. (2.12) has an exponential dichotomy.

Now we assume that the equation has an exponential dichotomy and we construct explicitly a function V satisfying conditions 1 and 2. Given $t \in I$

and $\varphi \in C$, write φ in the form $\varphi = x + y$, where $x \in E(t)$ and $y \in F(t)$. Now let

$$V(t, \varphi) = -V_E(t, x) + V_F(t, y),$$

where

$$V_E(t, x) = \sup\{\|T(r, t)x\|e^{\lambda(r-t)} : r \geq t\}$$

and

$$V_F(t, y) = \sup\{\|T(r, t)y\|e^{\lambda(t-r)} : r \leq t\}.$$

Condition 1 holds automatically. To establish condition 2, first note that if $\varphi \in E(\tau)$ and $t \geq \tau$, then

$$
\begin{aligned}
|V(t, T(t, \tau)\varphi)| &= V_E(t, T(t, \tau)\varphi) \\
&= e^{-\lambda(t-\tau)} \sup\{\|T(r, \tau)\varphi\|e^{\lambda(r-\tau)} : r \geq t\} \\
&\leq e^{-\lambda(t-\tau)} V_E(\tau, \varphi)
\end{aligned}
$$

and that if $\varphi \in F(\tau)$ and $t \geq \tau$, then

$$
\begin{aligned}
V(t, T(t, \tau)\varphi) &= V_F(t, T(t, \tau)\varphi) \\
&= e^{\lambda(t-\tau)} \sup\{\|T(r, \tau)\varphi\|e^{\lambda(\tau-r)} : r \leq t\} \\
&\geq e^{\lambda(t-\tau)} V_F(\tau, \varphi).
\end{aligned}
$$

This establishes property (3.13).

Finally, we establish property (3.12). By definition we have

$$|V(t, \varphi)| = V_E(t, \varphi) \geq \|\varphi\| \quad \text{for } \varphi \in E(t)$$

and

$$|V(t, \varphi)| = V_F(t, \varphi) \geq \|\varphi\| \quad \text{for } \varphi \in F(t).$$

On the other hand, by (3.5) with $t = s$, we obtain

$$|V(t, \varphi)| = |V_E(t, \varphi)| \leq D'\|\varphi\| \quad \text{for } \varphi \in E(t)$$

and

$$|V(t, \varphi)| = V_F(t, \varphi) \leq D'\|\varphi\| \quad \text{for } \varphi \in F(t).$$

This completes the proof of the theorem. $\qquad\square$

3.3 Decompositions and Upper Bounds

Assuming that Eq. (2.12) has an exponential dichotomy on an interval $I \subset \mathbb{R}$ containing $\mathbb{R}_0^+$, we show in this section that the extension $T_0(t, s)$ of the associated evolution family $T(t, s)$ to the space C_0 behaves as if it also had an exponential dichotomy on I. This amounts to decompose the linear operator X_0 in (2.32) into what could be considered its stable and unstable components. As an application, in Sec. 3.4 we describe the projections of the variation of constants formula onto the stable and unstable spaces.

We start by decomposing the operator X_0. Assume that Eq. (2.12) has an exponential dichotomy on an interval I containing $\mathbb{R}_0^+$. For each $t \in I$ we define linear operators $P_0(t), Q_0(t) \colon \mathbb{R}^n \to C_0$ by

$$P_0(t) = X_0 - Q_0(t)$$

and

$$Q_0(t) = \overline{T}(t, t + r)Q(t + r)T_0(t + r, t)X_0.$$

Clearly,

$$P_0(t) + Q_0(t) = X_0. \tag{3.14}$$

By construction, for each $p \in \mathbb{R}^n$ we have

$$P_0(t)p \in C_0 \setminus C \quad \text{and} \quad Q_0(t)p \in F(t) \subset C. \tag{3.15}$$

Indeed, by (2.28), the function $\varphi = T_0(t + r, t)X_0 p$ is continuous and so

$$Q_0(t)p = \overline{T}(t + r, t)Q(t + r)\varphi$$

is well defined and belongs to $F(t)$.

Theorem 3.4 (Hyperbolicity on C_0). *Assume that* (2.22) *holds and that Eq.* (2.12) *has an exponential dichotomy on an interval I containing $\mathbb{R}_0^+$. Then there exist constants $\lambda, N > 0$ such that*

$$\|T_0(t, s)P_0(s)\| \le Ne^{-\lambda(t-s)}, \quad \|\overline{T}(s, t)Q_0(t)\| \le Ne^{-\lambda(t-s)} \tag{3.16}$$

for every $t, s \in I$ with $t \ge s$.

Proof. Take $p \in \mathbb{R}^n$. Then

$$\|\overline{T}(s, t)Q_0(t)p\| = \|\overline{T}(s, t + r)Q(t + r)T_0(t + r, t)X_0 p\|$$
$$\le De^{-\lambda(t+r-s)}\|T_0(t + r, t)X_0 p\|$$

for $t \ge s$. On the other hand, by Proposition 2.4, we have

$$\|T_0(t + r, t)X_0 p\| \le Ke^{\omega r}\|X_0 p\| = K\|p\|e^{\omega r} \tag{3.17}$$

and so

$$\|\overline{T}(s,t)Q_0(t)p\| \le De^{-\lambda(t+r-s)}K\|p\|e^{\omega r}. \tag{3.18}$$

This establishes the second inequality in (3.16) taking $N \ge DKe^{(\omega-\lambda)r}$.

For the other inequality, first note that taking $t = s$ in (3.18) we obtain

$$\|Q_0(t)\| \le DKe^{(\omega-\lambda)r} \quad \text{for } t \in I.$$

Hence, for $t \in [s, s+r]$, using Proposition 2.4 we find that

$$\begin{aligned}
\|T_0(t,s)P_0(s)\| &\le Ke^{\omega r}\|X_0 - Q_0(s)\| \\
&\le Ke^{\omega r}(1 + N'),
\end{aligned} \tag{3.19}$$

where $N' = DKe^{(\omega-\lambda)r}$. Now take $t \ge s + r$. We have

$$\begin{aligned}
T_0(t,s)&P_0(s) \\
&= T_0(t,s)(X_0 - Q_0(s)) \\
&= T(t,s+r)T_0(s+r,s)(X_0 - Q_0(s)) \\
&= T(t,s+r)\big(T_0(s+r,s)X_0 - Q(s+r)T_0(s+r,s)X_0\big) \\
&= T(t,s+r)P(s+r)T_0(s+r,s)X_0.
\end{aligned} \tag{3.20}$$

Hence, by (3.4) and (3.17), we obtain

$$\|T_0(t,s)P_0(s)\| \le De^{-\lambda(t-s-r)}Ke^{\omega r}. \tag{3.21}$$

It follows from (3.19) and (3.21) that the first inequality in (3.16) holds taking a constant

$$N \ge Ke^{(\omega+\lambda)r}\max\{1 + N', D\}.$$

This completes the proof of the theorem. $\qquad\square$

In the following result we summarize a few properties of the evolution family when Eq. (2.12) has an exponential dichotomy.

Theorem 3.5. *Assume that condition (2.22) holds and that Eq. (2.12) has an exponential dichotomy on an interval I containing $\mathbb{R}_0^+$. Then there exist constants $\omega, K, \lambda, N > 0$ such that for each $t, s \in I$ with $t \ge s$ we have*

$$\|T(t,s)\| \le Ke^{\omega(t-s)}, \quad \|T_0(t,s)\| \le Ke^{\omega(t-s)}, \tag{3.22}$$

$$\|T(t,s)P(s)\| \le Ne^{-\lambda(t-s)}, \quad \|\overline{T}(s,t)Q(t)\| \le Ne^{-\lambda(t-s)}, \tag{3.23}$$

$$\|T(t,s)|E(s)\| \le Ne^{-\lambda(t-s)}, \quad \|\overline{T}(s,t)\| \le Ne^{-\lambda(t-s)},$$

$$\|T_0(t,s)P_0(s)\| \le Ne^{-\lambda(t-s)}, \quad \|\overline{T}(s,t)Q_0(t)\| \le Ne^{-\lambda(t-s)}. \tag{3.24}$$

Proof. The inequalities in (3.22) are established in Propositions 2.3 and 2.4. The remaining upper bounds in the theorem are given by (3.4) and Theorem 3.4. □

In particular, taking $t = s$ in (3.23) and (3.24) we obtain

$$\|P(t)\| \leq N, \quad \|Q(t)\| \leq N, \quad \|P_0(t)\| \leq N, \quad \|Q_0(t)\| \leq N$$

for every $t \in I$. In the remainder of the book we always use the constants in Theorem 3.5.

3.4 Projected Variation of Constants Formula

In this section we discuss how the variation of constants formula projects onto the stable and unstable spaces. We consider the perturbed Eq. (2.31), that is,

$$v' = L(t)v_t + g(t),$$

with the standing assumptions in the beginning of Sec. 2.2 and where $g \colon \mathbb{R} \to \mathbb{R}^n$ is a locally integrable function. Namely, we assume that:

(1) $L(t) \colon C \to \mathbb{R}^n$, for $t \in \mathbb{R}$, are bounded linear operators such that the map $(t, v) \mapsto L(t)v$ satisfies the Carathéodory conditions on $\mathbb{R} \times C$;

(2) the functions Π in (2.11) and $g \colon \mathbb{R} \to \mathbb{R}^n$ are locally integrable.

Theorem 3.6. *Assume that Eq. (2.12) has an exponential dichotomy on an interval I containing $\mathbb{R}_0^+$. Given $(s, \varphi) \in \Omega$, the unique solution $v \colon [s - r, +\infty) \to \mathbb{R}^n$ of Eq. (2.31) with $v_s = \varphi$ satisfies*

$$P(t)v_t = T(t,s)P(s)\varphi + \int_s^t T_0(t,\tau)P_0(\tau)g(\tau)\,d\tau$$

and

$$Q(t)v_t = T(t,s)Q(s)\varphi + \int_s^t T(t,\tau)Q_0(\tau)g(\tau)\,d\tau$$

for all $t \geq s$.

Proof. It follows from the variation of constants formula in (2.33) and (3.2) that

$$P(t)v_t = T(t,s)P(s)\varphi + P(t)\int_s^t T_0(t,\tau)X_0 g(\tau)\,d\tau$$

and

$$Q(t)v_t = T(t,s)Q(s)\varphi + Q(t)\int_s^t T_0(t,\tau)X_0 g(\tau)\,d\tau$$

for $t \geq s$. Note that

$$\int_s^t T_0(t,\tau) X_0 g(\tau)\, d\tau = v_t - T(t,s)\varphi$$

is a continuous function and so the projections of the integrals are indeed well defined. Hence, the claim in the theorem is equivalent to require that

$$P(t) \int_s^t T_0(t,\tau) X_0 g(\tau)\, d\tau = \int_s^t T_0(t,\tau) P_0(\tau) g(\tau)\, d\tau \qquad (3.25)$$

and

$$Q(t) \int_s^t T_0(t,\tau) X_0 g(\tau)\, d\tau = \int_s^t T(t,\tau) Q_0(\tau) g(\tau)\, d\tau \qquad (3.26)$$

for $t \geq s$. By (3.15) we have $Q_0(\tau) g(\tau) \in F(\tau)$ and so

$$\int_s^t T(t,\tau) Q_0(\tau) g(\tau)\, d\tau \in F(t).$$

Hence, identities (3.25) and (3.26) are equivalent to the single identity

$$Q(t) \int_s^t T_0(t,\tau) P_0(\tau) g(\tau)\, d\tau = 0. \qquad (3.27)$$

Before proceeding we establish an auxiliary result.

Lemma 3.1. *For each $\sigma \geq t \geq s$ and $p \in \mathbb{R}^n$ we have*

$$T(\sigma,t) \int_s^t T_0(t,\tau) X_0 p\, d\tau = \int_s^t T_0(\sigma,\tau) X_0 p\, d\tau,$$

where

$$\left(\int_s^t T_0(\sigma,\tau) X_0 p\, d\tau \right)(\theta) = \int_s^{t+\theta} (T_0(\sigma,\tau) X_0 p)(\theta)\, d\tau.$$

Proof of the lemma. Consider the functions u and v defined by

$$u_\sigma = T(\sigma,t) \int_s^t T_0(t,\tau) X_0 p\, d\tau \quad \text{and} \quad v_\sigma = \int_s^t T_0(\sigma,\tau) X_0 p\, d\tau$$

for $\sigma \geq t$. We have

$$u(\sigma) = \left(T(\sigma,t) \int_s^t T_0(t,\tau) X_0 p\, d\tau \right)(0)$$

$$= u(t) + \int_t^\sigma L(\rho) u_\rho\, d\rho$$

and

$$v(\sigma) = \int_s^t (T_0(\sigma, \tau) X_0 p)(0) \, d\tau$$

$$= \int_s^t \left[(T_0(t, \tau) X_0 p)(0) + \int_t^\sigma L(\rho) T_0(\rho, \tau) X_0 p \, d\rho \right] d\tau$$

$$= v(t) + \int_t^\sigma L(\rho) \int_s^t T_0(\rho, \tau) X_0 p \, d\tau \, d\rho$$

$$= v(t) + \int_t^\sigma L(\rho) v_\rho \, d\rho,$$

which shows that both u and v are solutions of Eq. (2.12). Since

$$u_t = v_t = \int_s^t T_0(t, \tau) X_0 p \, d\tau,$$

it follows from the uniqueness of solutions that $u = v$. $\qquad\square$

We proceed with the proof of the theorem. By Lemma 3.1, for $\sigma \geq t$ we have

$$T(\sigma, t) \int_s^t T_0(t, \tau) X_0 g(\tau) \, d\tau = \int_s^t T_0(\sigma, \tau) X_0 g(\tau) \, d\tau$$

and so

$$T(\sigma, t) \int_s^t T_0(t, \tau) P_0(\tau) g(\tau) \, d\tau$$

$$= T(\sigma, t) \int_s^t T_0(t, \tau)(X_0 - Q_0(\tau)) g(\tau) \, d\tau$$

$$= T(\sigma, t) \int_s^t T_0(t, \tau) X_0 g(\tau) \, d\tau - T(\sigma, t) \int_s^t T(t, \tau) Q_0(\tau) g(\tau) \, d\tau$$

$$= \int_s^t T_0(\sigma, \tau) X_0 g(\tau) \, d\tau - \int_s^t T(\sigma, \tau) Q_0(\tau) g(\tau) \, d\tau$$

$$= \int_s^t T_0(\sigma, \tau)(X_0 - Q_0(\tau)) g(\tau) \, d\tau.$$

Hence, for $\sigma \geq t + r$ we obtain

$$T(\sigma, t) \int_s^t T_0(t, \tau) P_0(\tau) g(\tau) \, d\tau$$

$$= \int_s^t T_0(\sigma, \tau) P_0(\tau) g(\tau) \, d\tau$$

$$= \int_s^t T(\sigma, \tau + r) P(\tau + r) T_0(\tau + r, \tau) X_0 g(\tau) \, d\tau \in E(\sigma),$$

using (3.20) in the second equality. Therefore,

$$Q(\sigma)T(\sigma,t)\int_s^t T_0(t,\tau)P_0(\tau)g(\tau)\,d\tau = 0$$

and since $T(\sigma,t)|F(t)$ is an isomorphism from $F(t)$ onto $F(\sigma)$, we conclude that identity (3.27) holds. This completes the proof of the theorem. $\square$

We note that under the assumptions of Theorem 3.6, the unique solution $v\colon [s-r,+\infty) \to \mathbb{R}^n$ of Eq. (2.31) with $v_s = \varphi$ also satisfies

$$P(t)v_t = T(t,\overline{s})P(\overline{s})v_{\overline{s}} + \int_{\overline{s}}^t T_0(t,\tau)P_0(\tau)g(\tau)\,d\tau$$

and

$$Q(t)v_t = T(t,\overline{s})Q(\overline{s})v_{\overline{s}} + \int_{\overline{s}}^t T(t,\tau)Q_0(\tau)g(\tau)\,d\tau \qquad (3.28)$$

for all $t \geq \overline{s} \geq s$. Moreover, since the linear operator $\overline{T}(t,s)$ in (3.3) is invertible, it follows readily from (3.28) that

$$Q(\overline{s})v_{\overline{s}} = \overline{T}(\overline{s},t)Q(t)v_t - \int_{\overline{s}}^t \overline{T}(\overline{s},\tau)Q_0(\tau)g(\tau)\,d\tau$$

for all $t \geq \overline{s} \geq s$.

3.5 The Autonomous Case

In this section we consider briefly the *autonomous* case, with the main purpose of illustrating the notion of an exponential dichotomy with a principal example.

Consider the equation $v' = Lv_t$, where $L\colon C \to \mathbb{R}^n$ is a bounded linear operator. We also consider the semigroup $S(t)$ introduced in (2.41).

Proposition 3.3. *$S(t)$ is strongly continuous, that is, $S(t)$ is a semigroup of bounded linear operators such that*

$$\lim_{t\searrow 0}\|S(t)\varphi - \varphi\| = 0 \quad \text{for } \varphi \in C. \qquad (3.29)$$

Proof. By (2.43) and (2.44), the family $S(t)$ is indeed a semigroup of bounded linear operators. Moreover, property (3.29) is a consequence of Proposition 2.2. Alternatively, let $v_t = S(t)\varphi$ and observe that

$$(S(t)\varphi - \varphi)(\theta) = v(t+\theta) - v(\theta).$$

Therefore,

$$\|S(t)\varphi - \varphi\| = \sup_{\theta\in[-r,0]} |v(t+\theta) - v(\theta)| \to 0$$

when $t \to 0$, in view of the uniform continuity of v on compact sets. $\square$

The *generator* $A \colon D(A) \to C$ of the strongly continuous semigroup $S(t)$ is defined by

$$A\varphi = \lim_{t \searrow 0} \frac{S(t)\varphi - \varphi}{t} \tag{3.30}$$

in the domain $D(A) \subset C$ of all $\varphi \in C$ for which the limit exists in the norm topology of C (one can show that this is equivalent to the differentiability of the map $t \mapsto S(t)\varphi$ on $\mathbb{R}^+$). The generator A is closed, densely defined, and determines the semigroup. Moreover, for each $\varphi \in D(A)$ we have

$$\frac{d}{dt}S(t)\varphi = AS(t)\varphi = S(t)A\varphi \tag{3.31}$$

for $t \geq 0$. See for example [103] for details and proofs of these properties.

Now we establish a few additional properties of the operator A. Writing L in the form (2.45), we define

$$\Delta(\lambda) = \int_{-r}^{0} e^{\lambda\theta} d\eta(\theta) - \lambda \mathrm{Id},$$

where Id is the $n \times n$ identity matrix.

Theorem 3.7. *Assume that $r > 0$. The following properties hold:*

(1) $A\varphi = \varphi'$ *on*

$$D(A) = \{\varphi \in C : \varphi' \in C, \varphi'(0) = L\varphi\};$$

(2) the spectrum $\sigma(A)$ of A is the set of all numbers $\lambda \in \mathbb{C}$ for which $A - \lambda\mathrm{Id}$ has no inverse and $\lambda \in \sigma(A)$ if and only if $\det \Delta(\lambda) = 0$;

(3) given $\gamma > 0$, there are finitely many numbers $\lambda \in \sigma(A)$ satisfying $\mathrm{Re}\,\lambda > \gamma$ and

$$\sup\{\mathrm{Re}\,\lambda : \lambda \in \sigma(A)\} < +\infty;$$

(4) the root space M_λ of each $\lambda \in \sigma(A)$ is finite-dimensional and there exists $k \in \mathbb{N}$ such that $M_\lambda = \mathrm{Ker}(A - \lambda\mathrm{Id})^k$.

Proof. For the first property, note that

$$(S(t)\varphi)(\theta) = \begin{cases} \varphi(t + \theta) & \text{if } t + \theta \leq 0, \\ \varphi(0) + \int_0^{t+\theta} LS(\tau)\varphi \, d\tau & \text{if } t + \theta > 0. \end{cases}$$

Hence, for $\theta < 0$ we have

$$\lim_{t \searrow 0} \frac{(S(t)\varphi)(\theta) - \varphi(\theta)}{t} = \varphi'(\theta). \tag{3.32}$$

On the other hand, for $\theta = 0$,

$$\lim_{t \searrow 0} \frac{(S(t)\varphi)(0) - \varphi(0)}{t} = \lim_{t \searrow 0} \frac{1}{t} \int_0^t LS(\tau)\varphi \, d\tau = L\varphi. \tag{3.33}$$

Since the limit in (3.30) exists in the norm topology of C, it also exists uniformly. Hence, $A\varphi$ must be a continuous function and so $\varphi'(0) = L\varphi$ (here $\varphi'(0)$ is the left-hand derivative since 0 is an endpoint of the domain). This shows that $A\varphi = \varphi'$ and

$$D(A) \subset \{\varphi \in C : \varphi' \in C, \varphi'(0) = L\varphi\}. \tag{3.34}$$

On the other hand, for a C^1 function φ with $\varphi'(0) = L\varphi$, it follows from (3.32) and (3.33) that $A\varphi = \varphi'$ and so the inclusion in (3.34) is an equality.

For the second property, we start by observing that by property 1, given $\varphi \in C$, the equation $(A - \lambda \mathrm{Id})\psi = \varphi$ has a solution if and only if

$$\psi' - \lambda\psi = \varphi \quad \text{and} \quad \psi'(0) = L\psi.$$

We obtain

$$\psi(\theta) = e^{\lambda\theta}\psi(0) + \int_\theta^0 e^{\lambda(\theta-\tau)}\varphi(\tau) \, d\tau$$

for $\theta \in [-r, 0]$ and so the condition $\psi'(0) = L\psi$ becomes

$$\lambda\psi(0) - \varphi(0) = \int_{-r}^0 d\eta(\theta)e^{\lambda\theta}\psi(0) + \int_{-r}^0 d\eta(\theta) \int_\theta^0 e^{\lambda(\theta-\tau)}\varphi(\tau) \, d\tau,$$

that is,

$$-\Delta(\lambda)\psi(0) = \varphi(0) + \int_{-r}^0 d\eta(\theta) \int_\theta^0 e^{\lambda(\theta-\tau)}\varphi(\tau) \, d\tau.$$

Therefore, when $\det \Delta(\lambda) \neq 0$ one can always find $\psi(0)$ for a given φ and so the operator $A - \lambda \mathrm{Id}$ has an inverse. This shows that

$$\sigma(A) \subset \{\lambda \in \mathbb{C} : \det \Delta(\lambda) = 0\}.$$

Now take $\lambda \in \mathbb{C}$ with $\det \Delta(\lambda) = 0$. Moreover, take $\varphi(\theta) = e^{\lambda\theta}x$ with $x \in \mathbb{R}^n \setminus \{0\}$ and $\Delta(\lambda)x = 0$. Then

$$L\varphi - \varphi'(0) = \int_{-r}^0 d\eta(\theta)e^{\lambda\theta}x - \lambda x$$
$$= \Delta(\lambda)x = 0$$

and so $\varphi \in D(A)$. Furthermore, $A\varphi = \lambda\varphi$ and λ belongs to the point spectrum of A. This establishes property 2.

For the last properties in the theorem, we first show that $S(t)$ is compact for $t \geq r$. The argument is a variation of the proof of Theorem 2.5. Consider the ball

$$B = \{\varphi \in C : \|\varphi\| \leq \rho\}$$

and take $\varphi \in B$ and $t \geq r$. We have

$$\|S(t)\varphi\| \leq e^{\|L\|t}\|\varphi\| < e^{\|L\|t}\rho.$$

Moreover, since $\psi = S(t)\varphi$ is of class C^1 for $t \geq r$ with

$$
\begin{aligned}
\psi'(\theta) &= \frac{d}{d\theta}(S(t)\varphi)(\theta) \\
&= \frac{d}{d\theta}(S(t+\theta)\varphi)(0) \\
&= \frac{d}{dt}(S(t+\theta)\varphi)(0) \\
&= L(S(t+\theta)\varphi),
\end{aligned}
$$

we obtain

$$\|\psi'\| \leq \|L\|e^{\|L\|t}\|\varphi\| \leq \|L\|e^{\|L\|t}\rho.$$

Hence,

$$S(t)B \subset \left\{\psi \in C : \psi' \in C, \|\psi\| \leq e^{\|L\|t}, \|\psi'\| \leq \|L\|e^{\|L\|t}\rho\right\} \subset B',$$

where

$$B' = \left\{\psi \in C : \|\psi\| \leq e^{\|L\|t}, |\psi(\theta) - \psi(\overline{\theta})| \leq \rho'|\theta - \overline{\theta}| \text{ for } \theta, \overline{\theta} \in [-r, 0]\right\}$$

and $\rho' = \|L\|e^{\|L\|t}\rho$. In view of the Arzelà–Ascoli theorem, the set B' is compact. Therefore, the linear operator $S(t)$ is compact. The remaining properties follow now from standard results in the theory of semigroups (see for example [60]). □

We note that each set M_λ is a subspace of the complexification $C \oplus iC$ (in general this is unavoidable since λ need not be real).

Finally, we give a condition for the hyperbolicity of the semigroup $S(t)$.

Theorem 3.8. *If the spectrum $\sigma(A)$ does not intersect the imaginary axis, then the equation $v' = Lv_t$ has an exponential dichotomy on $\mathbb{R}$.*

Proof. We first show that the semigroup $S(t)$ can be extended to a group on each space M_λ. Let

$$\Phi_\lambda = \{\varphi_1^\lambda, \ldots, \varphi_d^\lambda\}$$

be a basis for the finite-dimensional space M_λ. Since $AM_\lambda \subset M_\lambda$, there exists a $d \times d$ matrix B_λ such that $A\Phi_\lambda = \Phi_\lambda B_\lambda$. Moreover, since $A\varphi = \varphi'$ and this identity extends naturally to the complexification of A, we obtain $\Phi_\lambda' = \Phi_\lambda B_\lambda$, leading to

$$\Phi_\lambda(\theta) = \Phi_\lambda(0)e^{B_\lambda \theta} \quad \text{for } \theta \in [-r, 0].$$

Moreover, by (3.31) we have

$$S(t)\Phi_\lambda = \Phi_\lambda e^{Bt} \quad \text{for } t \geq 0$$

and so

$$(S(t)\Phi_\lambda)(\theta) = \Phi_\lambda(0)e^{B_\lambda(t+\theta)}$$

for $t \geq 0$ and $\theta \in [-r, 0]$. This formula can be used to define $S(t)$ for $t < 0$.

Moreover, for each $\lambda \in \sigma(A)$ we have the decomposition

$$C \oplus iC = M_\lambda \oplus N_\lambda, \tag{3.35}$$

where

$$N_\lambda = \text{Im}(A - \lambda \text{Id})^k.$$

Since the semigroup $S(t)$ extends to a group on M_λ, we have an isomorphism $S(t)|M_\lambda \colon M_\lambda \to M_\lambda$ and so, in particular,

$$S(t)M_\lambda = M_\lambda. \tag{3.36}$$

On the other hand, by (3.31) we have $S(t)A\varphi = AS(t)\varphi$ and hence,

$$S(t)N_\lambda \subset N_\lambda. \tag{3.37}$$

Now we consider the spaces

$$E = C \cap \bigcap_{\text{Re } \lambda > 0} N_\lambda \quad \text{and} \quad F = C \cap \bigoplus_{\text{Re } \lambda > 0} M_\lambda.$$

In view of (3.35) we have

$$C = E \oplus F. \tag{3.38}$$

Let $P(t) = P$ and $Q(t) = \text{Id} - P$ be the projections (independent of t) associated with the decomposition in (3.38). We must show that

$$PS(t) = S(t)P \quad \text{for } t \geq 0. \tag{3.39}$$

It follows from (3.36) and (3.37) that

$$S(t)E \subset E \quad \text{and} \quad S(t)F = F. \tag{3.40}$$

Now write

$$S(t) = S(t)P + S(t)(\mathrm{Id} - P).$$

By (3.40), for each $\varphi \in C$ we have

$$S(t)P\varphi \in E \quad \text{and} \quad S(t)(\mathrm{Id} - P)\varphi \in F.$$

Hence, $PS(t)\varphi = S(t)P\varphi$, which establishes property (3.39).

The bounds in the notion of an exponential dichotomy follow now from Theorem 3.7. Indeed, since the semigroup $S(t)$ is strongly continuous, the point spectrum of $S(t)$ is $e^{t\sigma(A)}$ or $e^{t\sigma(A)} \cup \{0\}$ (see for example [60]). Hence, if $\mu \neq 0$ is in the point spectrum of $S(t)$ for some t, then there exists $\lambda \in \sigma(A)$ such that $e^{\lambda t} = \mu$. In addition, we have

$$\mathrm{Ker}(\mu\mathrm{Id} - S(t))^k = \mathrm{span}\{\mathrm{Ker}(\lambda\mathrm{Id} - A)^k : \lambda \in \sigma(A), e^{\lambda t} = \mu\}.$$

On the other hand, in view of properties 3 and 4 in Theorem 3.7 the intersection and direct sum

$$\bigcap_{\mathrm{Re}\,\lambda > 0} N_\lambda \quad \text{and} \quad \bigoplus_{\mathrm{Re}\,\lambda > 0} M_\lambda$$

are taken over finitely many numbers λ. Hence, by the former observations the exponential bounds for A on E and F (recall that $\sigma(A)$ does not intersect the imaginary axis) transfer readily to corresponding exponential bounds for $S(t)$ on E and F. This completes the proof of the theorem. $\square$

We end this section noting that the behavior in Theorem 3.8 for an exponential dichotomy in the autonomous case can be imitated easily in the nonautonomous case. Namely, consider a decomposition $C = E \oplus F$ and write $\varphi \in C$ in the form $\varphi = x + y$ with $x \in E$ and $y \in F$. Now consider the equation

$$(v', w') = (L(t)v_t, M(t)w_t), \tag{3.41}$$

for some linear operators $L(t), M(t) \colon F \to \mathbb{R}^n$ with

$$M(t)\varphi = A(t)\varphi(0).$$

Notice that $w' = M(t)w_t$ is an ordinary differential equation. Hence, the second component of the associated evolution family is invertible and if the equations

$$v' = L(t)v_t \quad \text{and} \quad w' = M(t)w_t,$$

have respectively an exponential contraction and an exponential expansion (that is, an exponential contraction when the time is reversed), then Eq. (3.41) has an exponential dichotomy.

PART II
Linear Stability

Chapter 4

Two-Sided Robustness

In this chapter we establish the robustness of an exponential dichotomy for a linear delay equation, in the sense that the existence of an exponential dichotomy persists under sufficiently small linear perturbations. We consider the general case of a noninvertible dynamics, which causes that we also need to establish the invertibility of the perturbed dynamics along its unstable direction. Our approach exhibits in a more or less explicit manner the new stable and unstable spaces. The first step is the construction of bounded solutions into the future and into the past, already with a prescribed exponential rate. These solutions are then used to construct stable and unstable spaces for the perturbed equation, which is quite natural since the stable and unstable spaces essentially correspond to the bounded solutions respectively into the future and into the past. To complete the proof of the robustness property we also obtain a bound for the norms of the projections onto the stable and unstable spaces of the perturbed equation. We emphasize that in this chapter we consider only exponential dichotomies on the line. The robustness property also holds for exponential dichotomies on the half-line, although the proof requires other methods (see Chap. 6).

4.1 Robustness Property

In this section we formulate a result on the robustness of an exponential dichotomy for a linear delay equation. The proof depends on various constructions that are then described in the following sections.

We consider linear perturbations of Eq. (2.12) of the form

$$v' = (L(t) + M(t))v_t, \qquad (4.1)$$

with the standing assumptions in the beginning of Sec. 2.2 and with analogous assumptions for the perturbation. Namely, let $L(t), M(t) \colon C \to \mathbb{R}^n$, for $t \in \mathbb{R}$, be bounded linear operators such that:

(1) the functions $(t, v) \mapsto L(t)v$ and $(t, v) \mapsto M(t)v$ satisfy the Carathéodory conditions on $\mathbb{R} \times C$;

(2) the functions Π in (2.11) and $t \mapsto \|M(t)\|$ are locally integrable.

We always assume in this chapter that condition (2.22) holds, that is,

$$\sup_{t \in \mathbb{R}} \int_t^{t+1} \Pi(\tau) \, d\tau < +\infty.$$

By Theorem 2.4, Eq. (4.1) determines an evolution family $\widetilde{T}(t, s)$, for $t \geq s$, such that

$$v_t = \widetilde{T}(t, s)\varphi$$

for any solution $v \colon [s - r, +\infty) \to \mathbb{R}^n$ of Eq. (4.1) with $v_s = \varphi \in C$. Moreover, by the variation of constants formula in Theorem 2.7 each solution v of Eq. (4.1) with $v_s = \varphi$ satisfies

$$v_t = T(t, s)\varphi + \int_s^t T_0(t, \tau)X_0 M(\tau)v_\tau \, d\tau \tag{4.2}$$

for all $t \geq s$, where $T(t, s)$ is the evolution family associated with Eq. (2.12). By Theorem 3.6 we also have

$$v_t = T(t, s)\varphi + \int_s^t T_0(t, \tau)P_0(\tau)M(\tau)v_\tau \, d\tau + \int_s^t T(t, \tau)Q_0(\tau)M(\tau)v_\tau \, d\tau$$

for all $t \geq s$. Using only the linear operators $T(t, s)$ and $\widetilde{T}(t, s)$, these identities can be written, respectively, in the form

$$\widetilde{T}(t, s) = T(t, s) + \int_s^t T_0(t, \tau)X_0 M(\tau)\widetilde{T}(\tau, s) \, d\tau \tag{4.3}$$

and

$$\widetilde{T}(t, s) = T(t, s) + \int_s^t T_0(t, \tau)P_0(\tau)M(\tau)\widetilde{T}(\tau, s) \, d\tau$$

$$+ \int_s^t T(t, \tau)Q_0(\tau)M(\tau)\widetilde{T}(\tau, s) \, d\tau,$$

for all $t \geq s$.

The following result establishes the robustness of an exponential dichotomy on $\mathbb{R}$ for a linear delay equation.

Theorem 4.1 (Robustness property). *Assume that Eq. (2.12) has an exponential dichotomy on* $\mathbb{R}$ *and that there exists* $\delta > 0$ *such that*

$$\sup_{t \in \mathbb{R}} \int_t^{t+1} \|M(\tau)\| \, d\tau \leq \delta. \tag{4.4}$$

If δ *is sufficiently small, then Eq. (4.1) has an exponential dichotomy on* $\mathbb{R}$. *Moreover, the stable and unstable spaces for Eq. (4.1) are isomorphic, respectively, to the stable and unstable spaces* $E(t)$ *and* $F(t)$ *for Eq. (2.12).*

The proof of Theorem 4.1 is given in Secs. 4.2–4.5. Note that condition (4.4) holds for example when

$$\|M(t)\| \leq \delta \quad \text{for } t \in \mathbb{R}.$$

It also follows from the proof that for each $\varepsilon > 0$ there exists $\delta' > 0$ such that if $\delta < \delta'$ and λ is the exponent in Definition 3.2 for Eq. (2.12), then $\lambda - \varepsilon$ is an exponent in Definition 3.2 for Eq. (4.1).

A simple consequence of the last statement in Theorem 4.1 is the robustness of an exponential contraction on $\mathbb{R}$. For completeness we provide a simple direct proof.

Theorem 4.2. *Assume that Eq. (2.12) has an exponential contraction on* $\mathbb{R}$ *and that there exists* $\delta > 0$ *such that (4.4) holds. If* δ *is sufficiently small, then Eq. (4.1) has an exponential contraction on* $\mathbb{R}$.

Proof. It follows from (4.3) that

$$\|\widetilde{T}(t,s)\| \leq \|T(t,s)\| + \int_s^t \|T_0(t,\tau)\| \cdot \|M(\tau)\| \cdot \|\widetilde{T}(\tau,s)\| \, d\tau$$

$$\leq Ne^{-\lambda(t-s)} + N \int_s^t e^{-\lambda(t-\tau)} \|M(\tau)\| \cdot \|\widetilde{T}(\tau,s)\| \, d\tau$$

$$= Ne^{-\lambda(t-s)} + Ne^{-\lambda(t-s)} \int_s^t \|M(\tau)\| e^{\lambda(\tau-s)} \|\widetilde{T}(\tau,s)\| \, d\tau.$$

Let $\alpha(t) = e^{\lambda(t-s)} \|\widetilde{T}(t,s)\|$. Then

$$\alpha(t) \leq N + N \int_s^t \|M(\tau)\| \alpha(\tau) \, d\tau$$

and so, applying Gronwall's lemma to the function $\alpha(t)$ we obtain

$$\alpha(t) \leq N \exp\left(N \int_s^t \|M(\tau)\| \, d\tau \right)$$

for $t \geq s$. On the other hand, using (4.4) we have

$$\int_s^t \|M(\tau)\|\,d\tau \leq \sum_{m=0}^{\lfloor t-s \rfloor} \int_{s+m}^{s+m+1} \|M(\tau)\|\,d\tau \leq \delta(t-s+1).$$

Hence,

$$\|\widetilde{T}(t,s)\| \leq Ne^{N\delta}e^{-(\lambda-N\delta)(t-s)}$$

for $t \geq s$. This completes the proof of the theorem. □

Another consequence of Theorem 4.1 concerns the nonautonomous perturbations of an autonomous linear equation. Namely, consider Eq. (2.40), that is, $v' = Lv_t$, where $L\colon C \to \mathbb{R}^n$ is a bounded linear operator. Moreover, let $A\colon D(A) \to C$ be the generator of the strongly continuous semigroup $S(t)$ associated with Eq. (2.40) (see (3.30)).

Combining Theorems 3.8 and 4.1 yields the following result.

Theorem 4.3. *Assume that the spectrum $\sigma(A)$ does not intersect the imaginary axis and that there exists $\delta > 0$ such that (4.4) holds. If δ is sufficiently small, then the equation*

$$v' = (L + M(t))v_t$$

has an exponential dichotomy on $\mathbb{R}$.

4.2 Construction of Bounded Solutions

In this section we start the proof of Theorem 4.1. Since the stable and unstable spaces correspond to the bounded solutions respectively into the future and into the past, we first construct bounded solutions of Eq. (4.2) precisely into the future and into the past, already with an exponential rate.

First we introduce appropriate Banach spaces. In order to construct solutions into the future, let

$$I = \big\{(t,s) \in \mathbb{R} \times \mathbb{R} : t \geq s\big\}$$

and given $\varepsilon \in (0, \lambda/2)$, consider the Banach space

$$\mathcal{A} = \big\{X\colon I \to \mathcal{L}(C) \text{ continuous} : \|X\| < +\infty\big\}$$

equipped with the norm

$$\|X\| := \sup\big\{\|X(t,s)\|e^{(\lambda-\varepsilon)(t-s)} : (t,s) \in I\big\}.$$

Recall that $\mathcal{L}(C)$ is the set of bounded linear operators from C into itself.

Lemma 4.1. *Provided that δ is sufficiently small, there exists a unique $X \in \mathcal{A}$ such that*

$$X(t,s) = T(t,s)P(s) + \int_s^t T_0(t,\tau)P_0(\tau)M(\tau)X(\tau,s)\,d\tau$$

$$- \int_t^{+\infty} \overline{T}(t,\tau)Q_0(\tau)M(\tau)X(\tau,s)\,d\tau \tag{4.5}$$

for every $(t,s) \in I$. Moreover, given $s \in \mathbb{R}$ and $\varphi \in C$, the function $v_t = X(t,s)\varphi$, for $t \geq s$, is a solution of Eq. (4.2).

Proof. Consider the operator $U: \mathcal{A} \to \mathcal{A}$ defined by

$$(UX)(t,s) = T(t,s)P(s) + \int_s^t T_0(t,\tau)P_0(\tau)M(\tau)X(\tau,s)\,d\tau$$

$$- \int_t^{+\infty} \overline{T}(t,\tau)Q_0(\tau)M(\tau)X(\tau,s)\,d\tau$$

for each $X \in \mathcal{A}$. Before proceeding, we show that the improper integral is well defined. By Theorem 3.5 we have

$$\int_t^{+\infty} \|\overline{T}(t,\tau)Q_0(\tau)\| \cdot \|M(\tau)\| \cdot \|X(\tau,s)\|\,d\tau$$

$$= \sum_{m=0}^{\infty} \int_{t+m}^{t+m+1} \|\overline{T}(t,\tau)Q_0(\tau)\| \cdot \|M(\tau)\| \cdot \|X(\tau,s)\|\,d\tau$$

$$\leq \sum_{m=0}^{\infty} N\|X\| \int_{t+m}^{t+m+1} e^{-\lambda(\tau-t)+(-\lambda+\varepsilon)(\tau-s)}\|M(\tau)\|\,d\tau$$

$$= \sum_{m=0}^{\infty} N\|X\|e^{(-\lambda+\varepsilon)(t-s)} \int_{t+m}^{t+m+1} e^{(-2\lambda+\varepsilon)(\tau-t)}\|M(\tau)\|\,d\tau$$

$$\leq \sum_{m=0}^{\infty} N\|X\|e^{(-\lambda+\varepsilon)(t-s)}e^{(-2\lambda+\varepsilon)m} \int_{t+m}^{t+m+1} \|M(\tau)\|\,d\tau$$

$$\leq \frac{\delta N}{1-e^{-2\lambda+\varepsilon}}e^{(-\lambda+\varepsilon)(t-s)}\|X\| < +\infty. \tag{4.6}$$

On the other hand, since

$$\int_s^t \|T_0(t,\tau)P_0(\tau)\| \cdot \|M(\tau)\| \cdot \|X(\tau,s)\|\,d\tau$$

$$\leq \sum_{m=0}^{\lfloor t-s \rfloor} N\|X\| \int_{s+m}^{s+m+1} \chi_{[s,t]}(\tau)e^{-\lambda(t-\tau)+(-\lambda+\varepsilon)(\tau-s)}\|M(\tau)\|\,d\tau$$

$$\leq N\|X\|e^{(-\lambda+\varepsilon)(t-s)} \sum_{m=0}^{\lfloor t-s \rfloor} \int_{s+m}^{s+m+1} e^{-\varepsilon(t-\tau)}\|M(\tau)\|\,d\tau$$

$$\leq \delta N\|X\|e^{(-\lambda+\varepsilon)(t-s)} \sum_{m=0}^{\lfloor t-s \rfloor} e^{-\varepsilon(t-s-m)}$$

$$\leq \frac{\delta N}{1-e^{-\varepsilon}} e^{(-\lambda+\varepsilon)(t-s)}\|X\|,$$

we obtain

$$\|(UX)(t,s)\| \leq \|T(t,s)P(s)\| + \int_s^t \|T_0(t,\tau)P_0(\tau)\| \cdot \|M(\tau)\| \cdot \|X(\tau,s)\|\,d\tau$$

$$+ \int_t^{+\infty} \|\overline{T}(t,\tau)Q_0(\tau)\| \cdot \|M(\tau)\| \cdot \|X(\tau,s)\|\,d\tau$$

$$\leq Ne^{-\lambda(t-s)} + \delta ce^{(-\lambda+\varepsilon)(t-s)}\|X\|,$$

where

$$c = N\left(\frac{1}{1-e^{-\varepsilon}} + \frac{1}{1-e^{-2\lambda+\varepsilon}}\right). \tag{4.7}$$

This implies that

$$\|UX\| \leq N + \delta c\|X\| < +\infty \tag{4.8}$$

and we have a well-defined operator $U\colon \mathcal{A} \to \mathcal{A}$. Similarly,

$$\|UX_1 - UX_2\| \leq \delta c\|X_1 - X_2\|$$

for every $X_1, X_2 \in \mathcal{A}$. Thus, for any sufficiently small δ the operator U is a contraction and there exists a unique $X \in \mathcal{A}$ such that $UX = X$.

It remains to show that $v_t = X(t,s)\varphi$, for $t \geq s$, is a solution of Eq. (4.2). Indeed, by (4.5) we have

$$X(t,s) - T(t,s)X(s,s)$$

$$= T(t,s)P(s) + \int_s^t T_0(t,\tau)P_0(\tau)M(\tau)X(\tau,s)\,d\tau$$

$$- \int_t^{+\infty} \overline{T}(t,\tau)Q_0(\tau)M(\tau)X(\tau,s)\,d\tau$$

$$- T(t,s)\left(P(s) - \int_s^{+\infty} \overline{T}(s,\tau)Q_0(\tau)M(\tau)X(\tau,s)\,d\tau\right)$$

$$= \int_s^t T_0(t,\tau)P_0(\tau)M(\tau)X(\tau,s)\,d\tau - \int_t^{+\infty} \overline{T}(t,\tau)Q_0(\tau)M(\tau)X(\tau,s)\,d\tau$$

$$+ \int_s^{+\infty} \overline{T}(t,\tau)Q_0(\tau)M(\tau)X(\tau,s)\,d\tau$$

$$= \int_s^t T_0(t,\tau)P_0(\tau)M(\tau)X(\tau,s)\,d\tau + \int_s^t \overline{T}(t,\tau)Q_0(\tau)M(\tau)X(\tau,s)\,d\tau$$

$$= \int_s^t T_0(t,\tau)X_0 M(\tau)X(\tau,s)\,d\tau$$

for every $t \geq s$, in view of (3.14). $\qquad\qquad\qquad\qquad\qquad\square$

We also show that the bounded solutions constructed in Lemma 4.1 have a semigroup property.

Lemma 4.2. *Provided that δ is sufficiently small, we have*

$$X(t,\tau)X(\tau,s) = X(t,s) \quad for\ t \geq \tau \geq s.$$

Proof. By Theorem 3.6 and Lemma 4.1, we have

$$X(t,\tau)X(\tau,s) = T(t,s)P(s) + \int_s^\tau T_0(t,\sigma)P_0(\sigma)M(\sigma)X(\sigma,\tau)X(\tau,s)\,d\sigma$$

$$+ \int_\tau^t T_0(t,\sigma)P_0(\sigma)M(\sigma)X(\sigma,\tau)X(\tau,s)\,d\sigma$$

$$- \int_t^{+\infty} \overline{T}(t,\sigma)Q_0(\sigma)M(\sigma)X(\sigma,\tau)X(\tau,s)\,d\sigma$$

and thus,

$$X(t,\tau)X(\tau,s) - X(t,s)$$

$$= \int_s^t T_0(t,\sigma)P_0(\sigma)M(\sigma)\big[X(\sigma,\tau)X(\tau,s) - X(\sigma,s)\big]\,d\sigma \qquad (4.9)$$

$$- \int_t^{+\infty} \overline{T}(T,\sigma)Q_0(\sigma)M(\sigma)\big[X(\sigma,\tau)X(\tau,s) - X(\sigma,s)\big]\,d\sigma.$$

Now let

$$I_s = \big\{(t,\tau) \in \mathbb{R} \times \mathbb{R} : t \geq \tau \geq s\big\}$$

and consider the Banach space

$$\mathcal{A}_s = \big\{Z : I_s \to \mathcal{L}(C) \text{ continuous} : \|Z\|_s < +\infty\big\}$$

equipped with the norm

$$\|Z\|_s := \sup\big\{\|Z(t,\tau)\| : (t,\tau) \in I_s\big\}.$$

Letting

$$z(t,\tau) = X(t,\tau)X(\tau,s) - X(t,s)$$

for $t \geq \tau \geq s$ (with s fixed), it follows from (4.9) that $U_s z = z$, where the operator $U_s \colon \mathcal{A}_s \to \mathcal{A}_s$ is defined by

$$(U_s Z)(t, \tau) = \int_\tau^t T_0(t, \sigma) P_0(\sigma) M(\sigma) Z(\sigma, s) \, d\sigma$$

$$- \int_t^{+\infty} \overline{T}(t, \sigma) Q_0(\sigma) M(\sigma) Z(\sigma, s) \, d\sigma$$

for each $Z \in \mathcal{A}_s$ and $(t, \tau) \in I_s$. Since

$$\int_\tau^t \|T_0(t, \sigma) P_0(\sigma)\| \cdot \|M(\sigma)\| \cdot \|Z(\sigma, s)\| \, d\sigma$$

$$+ \int_t^{+\infty} \|\overline{T}(t, \sigma) Q_0(\sigma)\| \cdot \|M(\sigma)\| \cdot \|Z(\sigma, s)\| \, d\sigma$$

$$\leq \sum_{n=0}^{\lfloor t-\tau \rfloor} \int_{\tau+n}^{\tau+n+1} e^{-\lambda(t-\sigma)} \|M(\sigma)\| \, d\sigma \|Z\|_s$$

$$+ \sum_{n=0}^{\infty} \int_{t+n}^{t+n+1} N e^{-\lambda(\sigma-t)} \|M(\sigma)\| \, d\sigma \|Z\|_s$$

$$\leq \sum_{n=0}^{\lfloor t-\tau \rfloor} N e^{-\lambda(t-\tau-n)} \int_{\tau+n}^{\tau+n+1} \|M(\tau)\| \, d\sigma \|Z\|_s$$

$$+ \sum_{n=0}^{\infty} N e^{-\lambda n} \int_{t+n}^{t+n+1} \|M(\sigma)\| \, d\sigma \|Z\|_s$$

$$\leq \frac{\delta N}{1 - e^{-\lambda}} \|Z\|_s + \frac{\delta N}{1 - e^{-\lambda}} \|Z\|_s$$

$$= \frac{2\delta N}{1 - e^{-\lambda}} \|Z\|_s,$$

the point $(U_s Z)(t, \tau)$ is well defined and

$$\|U_s Z\|_s \leq \frac{2\delta N}{1 - e^{-\lambda}} \|Z\|_s < +\infty.$$

Hence, we obtain an operator $U_s \colon \mathcal{A}_s \to \mathcal{A}_s$.

For each $Z_1, Z_2 \in \mathcal{A}_s$ and $t \geq \tau$, we have

$$\|(U_s Z_1)(t, \tau) - (U_s Z_2)(t, \tau)\|$$

$$\leq \int_\tau^t \|T_0(t, \sigma) P_0(\sigma)\| \cdot \|M(\sigma)\| \cdot \|Z_1(\sigma, s) - Z_2(\sigma, s)\| \, d\sigma$$

$$+ \int_t^{+\infty} \|\overline{T}(t, \sigma) Q_0(\sigma)\| \cdot \|M(\sigma)\| \cdot \|Z_1(\sigma, s) - Z_2(\sigma, s)\| \, d\sigma$$

$$\leq \left(\frac{\delta N}{1 - e^{-\lambda}} + \frac{\delta N}{1 - e^{-\lambda}} \right) \|Z_1 - Z_2\|_s = \frac{2\delta N}{1 - e^{-\lambda}} \|Z_1 - Z_2\|_s$$

and thus,

$$\|U_s Z_1 - U_s Z_2\|_s \le \frac{2\delta N}{1 - e^{-\lambda}} \|Z_1 - Z_2\|_s.$$

For any sufficiently small δ the operator U_s is a contraction and so there exists a unique $Z \in \mathcal{A}_s$ with $U_s Z = Z$. Since $0 \in \mathcal{A}_s$ and $U_s 0 = 0$, we conclude that $Z = 0$.

Moreover, it follows from Lemma 4.1 that

$$\begin{aligned}
\|X(t,\tau)X(\tau,s)\| &\le \|X(t,\tau)\| \cdot \|X(\tau,s)\| \\
&\le \|X\|^2 e^{-(\lambda-\varepsilon)(t-\tau)} e^{-(\lambda-\varepsilon)(\tau-s)} \\
&= \|X\|^2 e^{-(\lambda-\varepsilon)(t-s)} \le \|X\|^2
\end{aligned}$$

and

$$\|X(t,s)\| \le \|X\| e^{-(\lambda-\varepsilon)(t-s)} \le \|X\|$$

for $t \ge \tau \ge s$. This shows that $z \in \mathcal{A}_s$. Since $U_s z = z$, it follows from the uniqueness of the fixed point of U_s that $z = 0$. $\qquad\square$

We also construct bounded solutions into the past. The argument is analogous to the one for the bounded solutions into the future. Let

$$J = \left\{ (t,s) \in \mathbb{R} \times \mathbb{R} : t \le s \right\}$$

and consider the Banach space

$$\mathcal{B} = \left\{ Y : J \to \mathcal{L}(C) \text{ continuous} : \|Y\| < +\infty \right\}$$

equipped with the norm

$$\|Y\| := \sup\left\{ \|Y(t,s)\| e^{(-\lambda+\varepsilon)(t-s)} : (t,s) \in J \right\}.$$

Lemma 4.3. *Provided that δ is sufficiently small, there exists a unique $Y \in \mathcal{B}$ such that*

$$\begin{aligned}
Y(t,s) = \overline{T}(t,s)Q(s) &+ \int_{-\infty}^{t} T_0(t,\tau)P_0(\tau)M(\tau)Y(\tau,s)\,d\tau \\
&- \int_{t}^{s} \overline{T}(t,\tau)Q_0(\tau)M(\tau)Y(\tau,s)\,d\tau
\end{aligned} \tag{4.10}$$

for every $(t,s) \in J$. Moreover, given $s \in \mathbb{R}$ and $\varphi \in C$, the function $v_t = Y(t,s)\varphi$, for $t \le s$, is a solution of Eq. (4.2).

Proof. Consider the operator $V \colon \mathcal{B} \to \mathcal{B}$ defined by

$$(VY)(t,s) = \overline{T}(t,s)Q(s) + \int_{-\infty}^{t} T_0(t,\tau)P_0(\tau)M(\tau)Y(\tau,s)\,d\tau$$

$$- \int_{t}^{s} \overline{T}(t,\tau)Q_0(\tau)M(\tau)Y(\tau,s)\,d\tau$$

for each $Y \in \mathcal{B}$. First we show that the improper integral is well defined. By Theorem 3.5 we have

$$\int_{-\infty}^{t} \|T_0(t,\tau)P_0(\tau)\| \cdot \|M(\tau)\| \cdot \|Y(\tau,s)\|\,d\tau$$

$$\leq Ne^{(\lambda-\varepsilon)(t-s)}\|Y\| \sum_{m=0}^{\infty} \int_{t-m-1}^{t-m} e^{(2\lambda-\varepsilon)(\tau-t)}\|M(\tau)\|\,d\tau \qquad (4.11)$$

$$\leq \frac{\delta N}{1-e^{-2\lambda+\varepsilon}} e^{(\lambda-\varepsilon)(t-s)}\|Y\|.$$

On the other hand, since

$$\int_{t}^{s} \|\overline{T}(t,\tau)Q_0(\tau)\| \cdot \|M(\tau)\| \cdot \|Y(\tau,s)\|\,d\tau$$

$$\leq Ne^{(\lambda-\varepsilon)(t-s)}\|Y\| \sum_{m=0}^{\lfloor s-t \rfloor} \int_{s-m-1}^{s-m} e^{\varepsilon(t-\tau)}\|M(\tau)\|\,d\tau$$

$$\leq \frac{\delta N}{1-e^{-\varepsilon}} e^{(\lambda-\varepsilon)(t-s)}\|Y\|,$$

we obtain

$$\|(VY)(t,s)\| \leq \|\overline{T}(t,s)Q(s)\|$$

$$+ \int_{-\infty}^{t} \|T_0(t,\tau)P_0(\tau)\| \cdot \|M(\tau)\| \cdot \|Y(\tau,s)\|\,d\tau$$

$$+ \int_{t}^{s} \|\overline{T}(t,\tau)Q_0(\tau)\| \cdot \|M(\tau)\| \cdot \|Y(\tau,s)\|\,d\tau$$

$$\leq Ne^{\lambda(t-s)} + \delta c e^{(\lambda-\varepsilon)(t-s)}\|Y\|,$$

with c as in (4.7). This implies that

$$\|VY\| \leq N + \delta c\|Y\| < +\infty \qquad (4.12)$$

and we have a well-defined operator $V \colon \mathcal{B} \to \mathcal{B}$. Similarly,

$$\|VY_1 - VY_2\| \leq \delta c\|Y_1 - Y_2\|$$

for every $Y_1, Y_2 \in \mathcal{B}$. Thus, for any sufficiently small δ the operator V is a contraction and there exists a unique $Y \in \mathcal{B}$ such that $VY = Y$.

For the last property in the lemma, we note that

$$
\begin{aligned}
Y(s,s) - T(s,t)Y(t,s) &= Q(s) + \int_{-\infty}^{s} T_0(s,\tau)P_0(\tau)M(\tau)Y(\tau,s)\,d\tau \\
&\quad - T(s,t)\overline{T}(t,s)Q(s) \\
&\quad - T(s,t)\left(\int_{-\infty}^{t} T_0(t,\tau)P_0(\tau)M(\tau)Y(\tau,s)\,d\tau \right. \\
&\quad \left. - \int_{t}^{s} \overline{T}(t,\tau)Q_0(\tau)M(\tau)Y(\tau,s)\,d\tau \right) \\
&= \int_{t}^{s} T_0(s,\tau)P_0(\tau)M(\tau)Y(\tau,s)\,d\tau \\
&\quad + \int_{t}^{s} \overline{T}(s,\tau)Q_0(\tau)M(\tau)Y(\tau,s)\,d\tau \\
&= \int_{t}^{s} T_0(s,\tau)X_0 M(\tau)Y(\tau,s)\,d\tau
\end{aligned}
$$

for each $t \leq s$. $\qquad\qquad\square$

We also establish a corresponding semigroup property for the solutions constructed in Lemma 4.3.

Lemma 4.4. *Provided that δ is sufficiently small, we have*

$$Y(t,\tau)Y(\tau,s) = Y(t,s) \quad for\ t \leq \tau \leq s.$$

Proof. To the possible extent, we follow the proof of Lemma 4.2. We have

$$
\begin{aligned}
Y(t,\tau)Y(\tau,s) &= \overline{T}(t,s)Q(s) - \int_{\tau}^{s} \overline{T}(t,\sigma)Q_0(\sigma)M(\sigma)Y(\sigma,s)\,d\sigma \\
&\quad + \int_{-\infty}^{t} T_0(t,\sigma)P_0(\sigma)M(\sigma)Y(\sigma,\tau)Y(\tau,s)\,d\sigma \\
&\quad - \int_{t}^{\tau} \overline{T}(t,\sigma)Q_0(\sigma)M(\sigma)Y(\sigma,\tau)Y(\tau,s)\,d\sigma
\end{aligned}
$$

and thus,

$$
\begin{aligned}
&Y(t,\tau)Y(\tau,s) - Y(t,s) \\
&= \int_{-\infty}^{t} T_0(t,\sigma)P_0(\sigma)M(\sigma)\big[Y(\sigma,\tau)Y(\tau,s) - Y(\sigma,s)\big]\,d\sigma \qquad (4.13) \\
&\quad - \int_{t}^{s} \overline{T}(t,\sigma)Q_0(\sigma)M(\sigma)\big[Y(\sigma,\tau)Y(\tau,s) - Y(\sigma,s)\big]\,d\sigma.
\end{aligned}
$$

Now let

$$J_s = \big\{(t,\tau) \in \mathbb{R} \times \mathbb{R} : t \leq \tau \leq s\big\}$$

and consider the Banach space

$$\mathcal{B}_s = \big\{ Z \colon J_s \to \mathcal{L}(C) \text{ continuous} : \|Z\|_s < +\infty \big\}$$

equipped with the norm

$$\|Z\|_s := \sup\big\{ \|Z(t,\tau)\| : (t,\tau) \in J_s \big\}.$$

Letting

$$z(t,s) = Y(t,\tau)Y(\tau,s) - Y(t,s)$$

for $t \le \tau \le s$ (with s fixed), it follows from (4.13) that $V_s z = z$, where the operator $V_s \colon \mathcal{B}_s \to \mathcal{B}_s$ is defined by

$$(V_s Z)(t,\tau) = \int_{-\infty}^{t} T_0(t,\sigma) P_0(\sigma) M(\sigma) Z(\sigma,\tau) \, d\sigma$$

$$- \int_{t}^{\tau} \overline{T}(t,\sigma) Q_0(\sigma) M(\sigma) Z(\sigma,\tau) \, d\sigma$$

for each $Z \in \mathcal{B}_s$ and $(t,\tau) \in J_s$. Proceeding as in the proof of Lemma 4.2, we find that 0 is the unique fixed point of V_s in $\mathcal{B}_s$. Since $z \in \mathcal{B}_s$ is also a fixed point, it follows that $z = 0$. $\square$

4.3 Solutions with Exponential Decay

In this section we show that all solutions of Eq. (4.2) exhibiting a certain exponential decay are those constructed in Lemmas 4.1 and 4.3.

We first consider the solutions into the future.

Lemma 4.5. *Given $(s,\varphi) \in \mathbb{R} \times C$, if $v \colon [s-r, +\infty) \to \mathbb{R}^n$ is a bounded solution of Eq. (4.2) with $v_s = \varphi$ such that*

$$\sup\big\{ \|v_t\| e^{(\lambda - \varepsilon)(t-s)} : t \ge s \big\} < +\infty,$$

then $v_t = X(t,s)\varphi$ for $t \ge s$ (for any sufficiently small δ).

Proof. For each $\tau \ge \sigma \ge s$ we have

$$P(\tau)v_\tau = T(\tau,\sigma) P(\sigma) v_\sigma + \int_{\sigma}^{\tau} T_0(\tau,u) P_0(u) M(u) v_u \, du \qquad (4.14)$$

and

$$Q(\tau)v_\tau = T(\tau,\sigma) Q(\sigma) v_\sigma + \int_{\sigma}^{\tau} T(\tau,u) Q_0(u) M(u) v_u \, du. \qquad (4.15)$$

Taking $\sigma = t$, we obtain

$$Q(t)v_t = \overline{T}(t,\tau) Q(\tau) v_\tau - \int_{t}^{\tau} \overline{T}(t,u) Q_0(u) M(u) v_u \, du. \qquad (4.16)$$

Since v is bounded, we have

$$\|\overline{T}(t,\tau)Q(\tau)v_\tau\| \le \ell N e^{-\lambda(\tau-t)}$$

for some constant $\ell > 0$. Finally, letting $\tau \to +\infty$ in (4.16) we obtain

$$Q(t)v_t = -\int_t^{+\infty} \overline{T}(t,u)Q_0(u)M(u)v_u \, du.$$

Adding this equation to (4.14) with $\tau = t$ and $\sigma = s$ yields the identity

$$v_t = T(t,s)P(s)\varphi + \int_s^t T_0(t,u)P_0(u)M(u)v_u \, du \qquad (4.17)$$
$$- \int_t^{+\infty} \overline{T}(t,u)Q_0(u)M(u)v_u \, du.$$

In other words, all bounded solutions into the future satisfy this identity.

For the statement in the lemma, we consider the Banach space

$$\mathcal{A}' = \{v \colon [s-r,+\infty) \to \mathbb{R}^n \text{ continuous} : \|v\| < +\infty\}$$

equipped with the norm

$$\|v\| := \sup\{\|v_t\|e^{(\lambda-\varepsilon)(t-s)} : t \ge s\}.$$

We also consider the operator $R \colon \mathcal{A}' \to \mathcal{A}'$ defined by

$$(Rv)_t = T(t,s)P(s)\varphi + \int_s^t T_0(t,u)P_0(u)M(u)v_u \, du$$
$$- \int_t^{+\infty} \overline{T}(t,u)Q_0(u)M(u)v_u \, du$$

for $v \in \mathcal{A}'$ and $t \ge s$. In a similar manner to that in the proof of Lemma 4.1, we find that

$$\|Rv_1 - Rv_2\| \le \delta c\|v_1 - v_2\|$$

for $v_1, v_2 \in \mathcal{A}'$. Hence, for any δ as in Lemma 4.1 the operator R is a contraction and so there exists a unique $v \in \mathcal{A}'$ such that $Rv = v$. By (4.17) this is the solution in the lemma.

Moreover, it follows from (4.5) that

$$X(t,s)\varphi = T(t,s)P(s)\varphi + \int_s^t T_0(t,u)P_0(u)M(u)X(u,s)\varphi \, du$$
$$- \int_t^{+\infty} \overline{T}(t,u)Q_0(u)M(u)X(u,s)\varphi \, du$$

for $t \ge s$, that is, the function $t \mapsto X(t,s)\varphi$ is a fixed point of R. We have

$$\sup\{\|X(t,s)\varphi\|e^{(\lambda-\varepsilon)(t-s)} : t \ge s\} \le \|X\| < +\infty$$

and so it follows from the uniqueness of the fixed point of R in the space $\mathcal{A}'$ that $v_t = X(t,s)\varphi$ for $t \ge s$. This completes the proof of the lemma. $\quad\square$

The exponential decay of the solutions into the past can be obtained as in the proof of Lemma 4.5.

Lemma 4.6. *Given $(s, \varphi) \in \mathbb{R} \times C$, if $v \colon (-\infty, s] \to \mathbb{R}^n$ is a bounded solution of Eq. (4.2) with $v_s = \varphi$ such that*

$$\sup\{\|v_t\|e^{(-\lambda+\varepsilon)(t-s)} : t \leq s\} < +\infty,$$

then $v_t = Y(t, s)\varphi$ for $t \leq s$ (for any sufficiently small δ).

Proof. Note that identities (4.14) and (4.15) hold for each $s \geq \tau \geq \sigma$. Moreover, the second identity also holds for $\tau \leq \sigma \leq s$ provided that T is replaced (twice) by $\overline{T}$, that is,

$$Q(\tau)v_\tau = \overline{T}(\tau, \sigma)Q(\sigma)v_\sigma - \int_\tau^\sigma \overline{T}(\tau, u)Q_0(u)M(u)v_u \, du \qquad (4.18)$$

for $\tau \leq \sigma \leq s$. Taking $\tau = t$, it follows from (4.14) that

$$P(t)v_t = T(t, \sigma)P(\sigma)v_\sigma + \int_\sigma^t T_0(t, u)P_0(u)g(u, v_u) \, du. \qquad (4.19)$$

Moreover, it follows from the boundedness of v that

$$\|\overline{T}(t, \sigma)P(\sigma)v_\sigma\| \leq \ell N e^{-\lambda(t-\sigma)}$$

for some constant $\ell > 0$. Hence, letting $\sigma \to -\infty$ in (4.19) we obtain

$$P(t)v_t = \int_{-\infty}^t T_0(t, u)P_0(u)g(u, v_u) \, du.$$

Finally, adding this equation to (4.18) with $\tau = t$ and $\sigma = s$ leads to

$$\begin{aligned} v_t = {} & \overline{T}(t, s)Q(s)\varphi + \int_{-\infty}^t T_0(t, u)P_0(u)M(u)v_u \, du \\ & - \int_t^s \overline{T}(t, u)Q_0(u)M(u)v_u \, du. \end{aligned} \qquad (4.20)$$

In other words, all bounded solutions into the past satisfy this identity.

For the statement in the lemma, we consider the Banach space

$$\mathcal{B}' = \{v \colon (-\infty, s] \to \mathbb{R}^n \text{ continuous} : \|v\| < +\infty\}$$

equipped with the norm

$$\|v\| := \sup\{\|v_t\|e^{(-\lambda+\varepsilon)(t-s)} : t \leq s\}.$$

We also consider the operator $S \colon \mathcal{B}' \to \mathcal{B}'$ defined by

$$\begin{aligned} (Sv)_t = {} & \overline{T}(t, s)Q(s)\varphi + \int_{-\infty}^t T_0(t, u)P_0(u)M(u)v_u \, du \\ & - \int_t^s \overline{T}(t, u)Q_0(u)M(u)v_u \, du. \end{aligned}$$

for each $v \in \mathcal{B}'$ and $t \leq s$. In a similar manner to that in the proof of Lemma 4.3, we obtain

$$\|Sv_1 - Sv_2\| \leq \delta c \|v_1 - v_2\|$$

for $v_1, v_2 \in \mathcal{B}'$. Hence, for δ as in Lemma 4.3 the operator S is a contraction and so there exists a unique $v \in \mathcal{B}'$ such that $Sv = v$. By (4.20) this is the solution in the lemma.

On the other hand, it follows from (4.10) that

$$Y(t,s)\varphi = \overline{T}(t,s)Q(s)\varphi + \int_{-\infty}^{t} T_0(t,u)P_0(u)M(u)Y(u,s)\varphi \, du$$
$$- \int_{t}^{s} \overline{T}(t,u)Q_0(u)M(u)Y(u,s)\varphi \, du$$

for $t \leq s$, that is, the function $t \mapsto Y(t,s)\varphi$ is a fixed point of S. Moreover,

$$\sup\{\|Y(t,s)\varphi\|e^{(-\lambda+\varepsilon)(t-s)} : t \leq s\} \leq \|Y\| < +\infty$$

and it follows from the uniqueness of the fixed point of S in the space $\mathcal{B}'$ that $v_t = Y(t,s)\varphi$ for $t \leq s$. This completes the proof of the lemma. $\square$

4.4 Construction of Projections

In this section we construct appropriate projections for the dynamics of the perturbed Eq. (4.2). Later on these will be used to establish the existence of an exponential dichotomy.

Observe that by Lemmas 4.2 and 4.4 the operators $X(t,t)$ and $Y(t,t)$ are projections. On the other hand, the stable and unstable spaces essentially correspond to the bounded solutions respectively into the future and into the past. This leads us to consider, for each $t \in \mathbb{R}$, the spaces

$$\widetilde{E}(t) = \operatorname{Im} X(t,t) \quad \text{and} \quad \widetilde{F}(t) = \operatorname{Im} Y(t,t).$$

Lemma 4.7. *Provided that δ is sufficiently small, for each $t \geq s$ we have*

$$\widetilde{T}(t,s)\widetilde{E}(s) \subset \widetilde{E}(t) \quad \text{and} \quad \widetilde{T}(t,s)\widetilde{F}(s) = \widetilde{F}(t).$$

Moreover, the map $\widetilde{T}(t,s)|\widetilde{F}(s) \colon \widetilde{F}(s) \to \widetilde{F}(t)$ is onto and invertible, with

$$\left(\widetilde{T}(t,s)|\widetilde{F}(s)\right)^{-1} = Y(s,t)|\widetilde{F}(t).$$

Proof. By Lemma 4.1, for each $\varphi \in C$ the function $v_t = X(t,s)\varphi$, for $t \geq s$, is a solution of Eq. (4.2) with $v_s = X(s,s)\varphi$. This implies that

$$X(t,s) = \widetilde{T}(t,s)X(s,s)$$

and by Lemma 4.2 we obtain

$$\widetilde{T}(t,s)\widetilde{E}(s) = \operatorname{Im} X(t,s)$$
$$= \operatorname{Im}\big(X(t,t)X(t,s)\big)$$
$$= X(t,t)\operatorname{Im} X(t,s) \subset \widetilde{E}(t)$$

for $t \geq s$. Analogously, by Lemma 4.3, the function $v_t = Y(t,s)\varphi$, for $t \leq s$, is a solution of Eq. (4.2), which implies that

$$Y(s,s) = \widetilde{T}(s,t)Y(t,s). \qquad (4.21)$$

Finally, by Lemma 4.4 we obtain

$$\widetilde{F}(s) = \widetilde{T}(s,t)\operatorname{Im} Y(t,s)$$
$$= \widetilde{T}(s,t)\operatorname{Im}\big(Y(t,t)Y(t,s)\big)$$
$$\subset \widetilde{T}(s,t)\widetilde{F}(t)$$

for $t \leq s$. Now we establish the reverse inclusion. Take $\varphi \in \widetilde{T}(s,t)\widetilde{F}(t)$. We must show that $\varphi \in \widetilde{F}(s)$ for each $t \leq s$. Take $\psi \in \widetilde{F}(t)$ such that $\widetilde{T}(s,t)\psi = \varphi$. Then $\tau \mapsto Y(\tau,t)\psi$, for $\tau \leq t$, is a bounded solution of Eq. (4.2) and so the function $v\colon (-\infty, s] \to \mathbb{R}^n$ defined by

$$v_\tau = \begin{cases} Y(\tau,t)\psi & \text{if } \tau \leq t, \\ \widetilde{T}(\tau,t)Y(t,t)\psi & \text{if } t \leq \tau \leq s \end{cases}$$

for $\tau \in (-\infty, s]$ is also a bounded solution of the same equation. Moreover,

$$\sup\big\{\|v_\tau\|e^{(-\lambda+\varepsilon)(\tau-s)} : \tau \leq s\big\} < +\infty.$$

It follows from Lemma 4.6 that there exists $\overline{\psi} \in C$ satisfying

$$Y(\tau,t)\psi = Y(\tau,s)\overline{\psi} \quad \text{for } \tau \leq t.$$

By (4.21), we obtain

$$\varphi = \widetilde{T}(s,t)Y(t,t)\psi$$
$$= \widetilde{T}(s,t)Y(t,s)\overline{\psi} = Y(s,s)\overline{\psi}.$$

Therefore, $\varphi \in \widetilde{F}(s)$ and so $\widetilde{T}(s,t)\widetilde{F}(t) = \widetilde{F}(s)$. For the last property in the lemma, take $\psi \in \widetilde{F}(t)$ and let $\overline{\psi}$ be as before. We have

$$\psi = Y(t,t)\psi = Y(t,s)\overline{\psi}$$
$$= Y(t,s)Y(s,s)\overline{\psi} \in \operatorname{Im}\big(Y(t,s)|\widetilde{F}(s)\big)$$

and so $\widetilde{F}(t) \subset \operatorname{Im}\big(Y(t,s)|\widetilde{F}(s)\big)$. Moreover,

$$\operatorname{Im}\big(Y(t,s)|\widetilde{F}(s)\big) = \operatorname{Im}\big(Y(t,t)Y(t,s)\big) \subset \widetilde{F}(t),$$

which leads to

$$\operatorname{Im}\big(Y(t,s)|\widetilde{F}(s)\big) = \widetilde{F}(t).$$

Finally, since $Y(s,s)^2 = Y(s,s)$, restricting (4.21) to $\widetilde{F}(s)$ we obtain

$$\operatorname{Id}_{\widetilde{F}(s)} = Y(s,s)|\widetilde{F}(s) = \widetilde{T}(s,t)Y(t,s)|\widetilde{F}(s).$$

This completes the proof of the lemma. □

By Lemma 4.7, we have maps

$$\widetilde{T}(t,s)|\widetilde{E}(s) = X(t,s)|\widetilde{E}(s)\colon \widetilde{E}(s) \to \widetilde{E}(t)$$

and

$$\big(\widetilde{T}(t,s)|\widetilde{F}(s)\big)^{-1} = Y(s,t)|\widetilde{F}(t)\colon \widetilde{F}(t) \to \widetilde{F}(s)$$

for $t \geq s$. Since $X \in \mathcal{A}$ and $Y \in \mathcal{B}$, we obtain

$$\big\|\widetilde{T}(t,s)|\widetilde{E}(s)\big\| \leq \eta e^{-(\lambda-\varepsilon)(t-s)} \tag{4.22}$$

and

$$\big\|\big(\widetilde{T}(t,s)|\widetilde{F}(s)\big)^{-1}\big\| \leq \eta e^{-(\lambda-\varepsilon)(t-s)} \tag{4.23}$$

for $t \geq s$, where $\eta = N/(1 - \delta c)$.

We use the results in the former lemmas to show that $\widetilde{E}(t)$ and $\widetilde{F}(t)$ form a direct sum. We start with an auxiliary statement about the operators

$$Z(t) = X(t,t) + Y(t,t).$$

Lemma 4.8. *Provided that δ is sufficiently small, the operator $Z(t)$ is invertible for every $t \in \mathbb{R}$.*

Proof. We have

$$Z(t) = P(t) - \int_t^{+\infty} \overline{T}(t,\tau)Q_0(\tau)M(\tau)X(\tau,t)\,d\tau$$

$$+ Q(t) + \int_{-\infty}^t T_0(t,\tau)P_0(\tau)M(\tau)Y(\tau,t)\,d\tau$$

and so

$$Z(t) - \operatorname{Id} = -\int_s^{+\infty} \overline{T}(t,\tau)Q_0(\tau)M(\tau)X(\tau,t)\,d\tau$$

$$+ \int_{-\infty}^t T_0(t,\tau)P_0(\tau)M(\tau)Y(\tau,t)\,d\tau.$$

By (4.6) and (4.11) we also have

$$\|Z(t) - \text{Id}\| \leq \int_t^{+\infty} \|\overline{T}(t,\tau)Q_0(\tau)\| \cdot \|M(\tau)\| \cdot \|X(\tau,t)\| \, d\tau$$
$$+ \int_{-\infty}^t \|T_0(t,\tau)P_0(\tau)\| \cdot \|M(\tau)\| \cdot \|Y(\tau,t)\| \, d\tau$$
$$\leq \frac{\delta N}{1 - e^{-2\lambda+\varepsilon}} (\|X\| + \|Y\|).$$

Moreover, by (4.8) and (4.12) we obtain

$$\|X\| \leq N/(1 - \delta c) \quad \text{and} \quad \|Y\| \leq N/(1 - \delta c). \tag{4.24}$$

Therefore, for any sufficiently small δ (independently of t), the operator $Z(t)$ is invertible. $\qquad\square$

Lemma 4.9. *Provided that δ is sufficiently small, we have $C = \widetilde{E}(t) \oplus \widetilde{F}(t)$ for $t \in \mathbb{R}$.*

Proof. Take $\varphi \in \widetilde{E}(t) \cap \widetilde{F}(t)$. By (4.22) and (4.23) we have

$$\frac{1}{\eta} e^{(\lambda-\varepsilon)(t-s)} \|\varphi\| \leq \|\widetilde{T}(t,s)\varphi\| \leq \eta e^{-(\lambda-\varepsilon)(t-s)} \|\varphi\|$$

for $t \geq s$. Since $\lambda > \varepsilon$, this leads to $\varphi = 0$. Hence,

$$\widetilde{E}(t) \cap \widetilde{F}(t) = \{0\}.$$

Moreover, by Lemma 4.8 the operator $Z(t)$ is invertible and so

$$C = Z(t)C \subset \text{Im}\, X(t,t) + \text{Im}\, Y(t,t) = \widetilde{E}(t) + \widetilde{F}(t).$$

This yields the desired statement. $\qquad\square$

It follows from Lemma 4.9 that given $t \in \mathbb{R}$, each $\varphi \in C$ can be written in a unique form $\varphi = x + y$ with $x \in \widetilde{E}(t)$ and $y \in \widetilde{F}(t)$. We define projections $\widetilde{P}(t), \widetilde{Q}(t) \colon C \to C$ such that $\widetilde{P}(t) + \widetilde{Q}(t) = \text{Id}$ by

$$\widetilde{P}(t)\varphi = x \quad \text{and} \quad \widetilde{Q}(t)\varphi = y. \tag{4.25}$$

Clearly,

$$\widetilde{E}(t) = \widetilde{P}(t)C \quad \text{and} \quad \widetilde{F}(t) = \widetilde{Q}(t)C.$$

Lemma 4.10. *Provided that δ is sufficiently small, we have*

$$\widetilde{P}(t)\widetilde{T}(t,s) = \widetilde{T}(t,s)\widetilde{P}(s) \quad \text{for } t \geq s. \tag{4.26}$$

Proof. Note that

$$\widetilde{T}(t,s) = \widetilde{T}(t,s)\widetilde{P}(s) + \widetilde{T}(t,s)\widetilde{Q}(s).$$

By Lemma 4.7, for each $\varphi \in C$ we have

$$\widetilde{T}(t,s)\widetilde{P}(s)\varphi \in \widetilde{E}(t) \quad \text{and} \quad \widetilde{T}(t,s)\widetilde{Q}(s)\varphi \in \widetilde{F}(t).$$

Therefore,

$$\widetilde{P}(t)\widetilde{T}(t,s)\varphi = \widetilde{T}(t,s)\widetilde{P}(s)\varphi$$

and

$$\widetilde{Q}(t)\widetilde{T}(t,s)\varphi = \widetilde{T}(t,s)\widetilde{Q}(s)\varphi,$$

which yields property (4.26). $\qquad\square$

4.5 Norms of the Projections

To complete the proof of Theorem 4.1 we need to obtain bounds for the norms of the projections $\widetilde{P}(t)$ and $\widetilde{Q}(t)$ in (4.25).

We first establish an auxiliary result. Following (3.6), let

$$\widetilde{\Theta}(t) = \inf\{\|w - z\| : w \in \widetilde{E}(t), z \in \widetilde{F}(t), \|w\| = \|z\| = 1\}.$$

Lemma 4.11. *Provided that δ is sufficiently small, there exists $\widetilde{\rho} > 0$ such that $\widetilde{\Theta}(t) \geq \widetilde{\rho}$ for $t \in \mathbb{R}$.*

Proof. Take vectors $w \in \widetilde{E}(t)$ and $z \in \widetilde{F}(t)$. Then there exist $x \in E(t)$ and $y \in F(t)$ such that

$$w = X(t,t)x = (\mathrm{Id} + G(t))x$$

and

$$z = Y(t,t)y = (\mathrm{Id} + H(t))y,$$

where

$$G(t) = -\int_t^{+\infty} \overline{T}(t,\tau)Q_0(\tau)M(\tau)X(\tau,t)\,d\tau$$

and

$$H(t) = \int_{-\infty}^t T_0(t,\tau)P_0(\tau)M(\tau)Y(\tau,t)\,d\tau.$$

It follows from (4.6) and (4.11) that

$$\|G(t)\| \leq \mu\|X\| \quad \text{and} \quad \|H(t)\| \leq \mu\|Y\|,$$

where

$$\mu = \frac{\delta N}{1 - e^{-2\lambda + \varepsilon}}.$$

Therefore,

$$\left(1 - \mu \|X\|\right)\|w\| \leq \|x\| \leq \left(1 + \mu \|X\|\right)\|x\| \tag{4.27}$$

and

$$\left(1 - \mu \|Y\|\right)\|z\| \leq \|y\| \leq \left(1 + \mu \|Y\|\right)\|y\|. \tag{4.28}$$

On the other hand, we have

$$\left\| \frac{x}{\|x\|} - \frac{y}{\|y\|} \right\| = \left\| \frac{(x-y)\|y\| + y(\|y\| - \|x\|)}{\|x\| \cdot \|y\|} \right\| \leq \frac{2}{\|x\|} \|x - y\|.$$

By (4.27) and (4.28), we obtain

$$\|w - z\| = \|x - y + G(t)x - H(t)y\|$$
$$\geq \|x - y\| - \|G(t)\| \cdot \|x\| - \|H(t)\| \cdot \|y\|$$
$$\geq \frac{\|x\|}{2} \left\| \frac{x}{\|x\|} - \frac{y}{\|y\|} \right\| - \frac{\|G(t)\|}{1 - \mu\|X\|} \|w\| - \frac{\|H(t)\|}{1 - \mu\|Y\|} \|z\|$$

and so

$$\|w - z\| \geq \frac{\|w\|}{2(1 + \mu\|X\|)} \left\| \frac{x}{\|x\|} - \frac{y}{\|y\|} \right\| - \frac{\mu\|X\|}{1 - \mu\|X\|} \|w\| - \frac{\mu\|Y\|}{1 - \mu\|Y\|} \|z\|.$$

Finally, taking the infimum over all w and z of norm 1 yields the inequality

$$\widetilde{\Theta}(t) \geq \frac{\Theta(t)}{2(1 + \mu\|X\|)} - \frac{\mu\|X\|}{1 - \mu\|X\|} - \frac{\mu\|Y\|}{1 - \mu\|Y\|}, \tag{4.29}$$

where

$$\Theta(t) = \inf\left\{ \|x - y\| : x \in E(t), y \in F(t), \|x\| = \|y\| = 1 \right\}.$$

On the other hand, it was shown in the proof of Theorem 3.1 (see (3.9)) that there exists a constant $\rho > 0$ such that

$$\Theta(t) \geq \rho \quad \text{for } t \in \mathbb{R}.$$

Hence, it follows from (4.24) and (4.29) that taking δ sufficiently small yields the desired property. $\qquad \square$

We proceed with the proof of Theorem 4.1. We already constructed projections $\widetilde{P}(t)$ such that (4.26) holds, which establishes condition 1 in Definition 3.2 for Eq. (4.2) and so also for Eq. (4.1). Moreover, by Lemma 4.7, condition 2 holds. To show that Eq. (4.1) has an exponential dichotomy, it remains to obtain the exponential bounds in condition 3 in Definition 3.2. In a similar to that in the proof of Theorem 3.1 (see (3.7)) we have

$$\|\widetilde{P}(t)\| \leq \frac{2}{\widetilde{\Theta}(t)} \quad \text{and} \quad \|\widetilde{Q}(t)\| \leq \frac{2}{\widetilde{\Theta}(t)}.$$

Therefore, it follows from Lemma 4.11 that

$$\|\widetilde{P}(t)\| \leq \frac{2}{\widetilde{\rho}} \quad \text{and} \quad \|\widetilde{Q}(t)\| \leq \frac{2}{\widetilde{\rho}} \tag{4.30}$$

for each $t \in \mathbb{R}$. Moreover, we have

$$\|\widetilde{T}(t,s)\widetilde{P}(s)\| \leq \|\widetilde{T}(t,s)|\widetilde{E}(s)\| \cdot \|\widetilde{P}(s)\|$$

and

$$\left\|\left(\widetilde{T}(t,s)|\widetilde{F}(s)\right)^{-1}\widetilde{Q}(t)\right\| \leq \left\|\left(\widetilde{T}(t,s)|\widetilde{F}(s)\right)^{-1}\right\| \cdot \|\widetilde{Q}(t)\|$$

for $t \geq s$. Therefore, the upper bounds in condition 3 follow from (4.30) together with (4.22) and (4.23). Summing up, Eq. (4.1) has an exponential dichotomy on $\mathbb{R}$.

For the last statement in the theorem we proceed as follows. By (4.5) we have

$$X(t,s)P(s) = T(t,s)P(s) + \int_s^t T_0(t,\tau)P_0(\tau)M(\tau)X(\tau,s)P(s)\,d\tau \\ - \int_t^{+\infty} \overline{T}(t,\tau)Q_0(\tau)M(\tau)X(\tau,s)P(s)\,d\tau \tag{4.31}$$

for $(t,s) \in I$. We also have

$$\|X(t,s)P(s)\|e^{(\lambda-\varepsilon)(t-s)} \leq N\|X\|$$

for $(t,s) \in I$, which shows that the function

$$I \ni (t,s) \mapsto X(t,s)P(s)$$

is in $\mathcal{A}$. By (4.31) this function is also a fixed point of the operator U in the proof of Lemma 4.1. In view of the uniqueness of the fixed point, we conclude that

$$X(t,s) = X(t,s)P(s) \quad \text{for } (t,s) \in I.$$

Analogously, by (4.10) we have

$$Y(t,s)Q(s) = \overline{T}(t,s)Q(s) + \int_{-\infty}^{t} T_0(t,\tau)P_0(\tau)M(\tau)Y(\tau,s)Q(s)\,d\tau$$
$$- \int_{t}^{s} \overline{T}(t,\tau)Q_0(\tau)M(\tau)Y(\tau,s)Q(s)\,d\tau \tag{4.32}$$

for $(t,s) \in J$. We also have

$$\|Y(t,s)Q(s)\|e^{(-\lambda+\varepsilon)(t-s)} \leq N\|Y\|$$

for $(t,s) \in J$, which shows that the function

$$J \ni (t,s) \mapsto Y(t,s)Q(s)$$

is in $\mathcal{B}$. By (4.32) this function is also a fixed point of the operator V in the proof of Lemma 4.3 and similarly,

$$Y(t,s) = Y(t,s)Q(s) \quad \text{for } (t,s) \in J.$$

On the other hand, by Lemma 4.8, the operator

$$Z(t) = X(t,t) + Y(t,t)$$
$$= X(t,t)P(t) + Y(t,t)Q(t)$$

is invertible. Since

$$\widetilde{E}(t) = \operatorname{Im} X(t,t)$$
$$= \operatorname{Im}\big(X(t,t)P(t)\big)$$
$$= Z(t)E(t)$$

and

$$\widetilde{F}(t) = \operatorname{Im} Y(t,t)$$
$$= \operatorname{Im}\big(Y(t,t)Q(t)\big)$$
$$= Z(t)F(t),$$

it follows from the invertibility of $Z(t)$ that the spaces $\widetilde{E}(t)$ and $\widetilde{F}(t)$ are isomorphic, respectively, to the spaces $E(t)$ and $F(t)$. This completes the proof of Theorem 4.1.

Chapter 5

Admissibility

In this chapter we consider the notion of admissibility and we show how it can be used to characterize the existence of an exponential dichotomy or of an exponential contraction for a linear delay equation. The notion of admissibility, essentially introduced by Perron, refers to the existence and sometimes uniqueness of solutions of a perturbed linear equation under bounded perturbations. After introducing appropriate Banach spaces where we take the perturbations and where we look for the solutions, we show that the existence of an exponential dichotomy is equivalent to the admissibility of those spaces. The major difficulty is showing that the admissibility property yields an exponential behavior, essentially due to the possible noninvertibility of the evolution families. We consider both delay equations on the line and on the half-line. The results in this chapter are used in Chap. 6 to establish the robustness of an exponential dichotomy and of an exponential contraction.

5.1 The Notion of Admissibility

In the following sections we characterize the existence of an exponential dichotomy on $\mathbb{R}$ for the linear equation $v' = L(t)v_t$ in terms of the admissibility of a certain pair of spaces $(\mathcal{E}, \mathcal{F})$. This refers to the existence and uniqueness of solutions $v \in \mathcal{E}$ of the perturbed equation

$$v' = L(t)v_t + g(t)$$

for all perturbations $g \in \mathcal{F}$. Here, $L(t) \colon C \to \mathbb{R}^n$, for $t \in \mathbb{R}$, are bounded linear operators such that:

(1) the map $(t, v) \mapsto L(t)v$ satisfies the Carathéodory conditions on $\mathbb{R} \times C$;
(2) the function Π in (2.11) satisfies condition (2.22).

The case of exponential dichotomies on $\mathbb{R}_0^+$, which requires a slightly different approach, is considered in Sec. 5.5.

We first introduce appropriate Banach spaces. Let $\mathcal{E}$ be the set of all continuous functions $v \colon \mathbb{R} \to \mathbb{R}^n$ such that

$$|v|_{\mathcal{E}} := \sup_{t \in \mathbb{R}} |v(t)| < +\infty$$

and let $\mathcal{F}$ be the set of all measurable functions $g \colon \mathbb{R} \to \mathbb{R}^n$ such that

$$|g|_{\mathcal{F}} := \sup_{t \in \mathbb{R}} \int_t^{t+1} |g(\tau)| \, d\tau < +\infty,$$

identified if they are equal almost everywhere (notice that each function $g \in \mathcal{F}$ is locally integrable). Note that $\mathcal{E}$ is a Banach space when equipped with the norm $|\cdot|_{\mathcal{E}}$.

Proposition 5.1. $\mathcal{F} = (\mathcal{F}, |\cdot|_{\mathcal{F}})$ *is a Banach space.*

Proof. Take a sequence $(g_m)_{m \in \mathbb{N}}$ in $\mathcal{F}$ such that

$$\sum_{m=1}^{\infty} |g_m|_{\mathcal{F}} < +\infty.$$

We define functions $f_m \colon \mathbb{R} \to \mathbb{R}_0^+$ for $m \in \mathbb{N}$ and $f \colon \mathbb{R} \to (-\infty, +\infty]$ by

$$f_m(t) = \sum_{p=1}^{m} |g_p(t)| \quad \text{and} \quad f(t) = \sum_{m=1}^{\infty} |g_m(t)|.$$

It follows from the monotone convergence theorem that

$$\int_t^{t+1} f(\tau) \, d\tau = \lim_{m \to \infty} \int_t^{t+1} f_m(\tau) \, d\tau$$

$$= \lim_{m \to \infty} \sum_{p=1}^{m} \int_t^{t+1} |g_p(\tau)| \, d\tau \le \sum_{m=1}^{\infty} |g_m|_{\mathcal{F}}$$

for $t \in \mathbb{R}$. Hence, $f(t) < +\infty$ for almost every $t \in \mathbb{R}$ and so the series $\sum_{m=1}^{\infty} g_m(t)$ converges for almost every $t \in \mathbb{R}$. Now let

$$g(t) = \sum_{m=1}^{\infty} g_m(t)$$

(which is defined almost everywhere). It follows from Fatou's lemma that

$$\int_t^{t+1} \sum_{p=1}^{\infty} |g_p(\tau)| \, d\tau = \int_t^{t+1} \lim_{m \to \infty} \sum_{p=1}^{m} |g_p(\tau)| \, d\tau$$

$$\le \liminf_{m \to \infty} \int_t^{t+1} \sum_{p=1}^{m} |g_p(\tau)| \, d\tau$$

$$= \liminf_{m \to \infty} \sum_{p=1}^{m} \int_t^{t+1} |g_p(\tau)| \, d\tau \le \sum_{m=1}^{\infty} |g_m|_{\mathcal{F}}$$

for all $t \in \mathbb{R}$ and thus, $g \in \mathcal{F}$. Moreover, again by Fatou's lemma, we have

$$
\int_t^{t+1} \left| g(\tau) - \sum_{p=1}^m g_p(\tau) \right| d\tau = \int_t^{t+1} \left| \sum_{p=m+1}^\infty g_m(\tau) \right| d\tau
$$

$$
\leq \sum_{p=m+1}^\infty \int_t^{t+1} |g_m(\tau)| \, d\tau
$$

$$
\leq \sum_{p=m+1}^\infty |g_m|_{\mathcal{F}}
$$

for all $t \in \mathbb{R}$. Therefore, the series $\sum_{m=1}^\infty g_m$ converges to g in $\mathcal{F}$, which implies that $\mathcal{F} = (\mathcal{F}, |\cdot|_{\mathcal{F}})$ is a Banach space. $\square$

Now we introduce the notion of admissibility used in the following sections for a linear delay equation on the line, that is, for $t \in \mathbb{R}$. Let $T(t, s)$ be the evolution family associated with Eq. (2.12) (see Definition 2.6) and let $T_0(t, s)$ be the linear operators introduced in Definition 2.10.

Definition 5.1. We say that the pair of spaces $(\mathcal{E}, \mathcal{F})$ is *admissible* for Eq. (2.12) if for each $g \in \mathcal{F}$ there exists a unique $v \in \mathcal{E}$ such that

$$
v_t = T(t, s)v_s + \int_s^t T_0(t, \tau) X_0 g(\tau) \, d\tau \tag{5.1}
$$

for $t, s \in \mathbb{R}$ with $t \geq s$.

In Sec. 5.5 we consider a slightly different notion of admissibility adapted to a linear delay equation on the half-line.

5.2 Hyperbolicity and Admissibility

We first show that the admissibility of the pair of spaces $(\mathcal{E}, \mathcal{F})$ is a consequence of the existence of an exponential dichotomy for Eq. (2.12).

Theorem 5.1 (Admissibility via hyperbolicity). *If Eq. (2.12) has an exponential dichotomy on $\mathbb{R}$, then the pair $(\mathcal{E}, \mathcal{F})$ is admissible for the equation.*

Proof. Given $g \in \mathcal{F}$, for each $t \in \mathbb{R}$ let $v_t = x_t + y_t$, where

$$
x_t = \int_{-\infty}^t T_0(t, \tau) P_0(\tau) g(\tau) \, d\tau
$$

and

$$y_t = - \int_t^{+\infty} T(t,\tau) Q_0(\tau) g(\tau) \, d\tau.$$

We first show that the integrals are well defined. By Theorem 3.5 we have

$$\int_{-\infty}^t \| T_0(t,\tau) P_0(\tau) g(\tau) \| \, d\tau \le \int_{-\infty}^t N e^{-\lambda(t-\tau)} |g(\tau)| \, d\tau$$

$$= \sum_{m=0}^\infty \int_{t-m-1}^{t-m} N e^{-\lambda(t-\tau)} |g(\tau)| \, d\tau$$

$$\le \sum_{m=0}^\infty N e^{-\lambda m} \int_{t-m-1}^{t-m} |g(\tau)| \, d\tau$$

$$\le \frac{N}{1-e^{-\lambda}} |g|_{\mathcal{F}}$$

and

$$\int_t^{+\infty} \| T(t,\tau) Q_0(\tau) g(\tau) \| \, d\tau \le \frac{N}{1-e^{-\lambda}} |g|_{\mathcal{F}},$$

for all $t \in \mathbb{R}$. Therefore, the continuous functions $x, y \colon \mathbb{R} \to \mathbb{R}^n$ satisfy

$$\sup_{t \in \mathbb{R}} |x(t)| < +\infty \quad \text{and} \quad \sup_{t \in \mathbb{R}} |y(t)| < +\infty,$$

and so $v \in \mathcal{E}$. Now take $s \in \mathbb{R}$. For $t \ge s$ we have

$$v_t = \int_s^t T_0(t,\tau) X_0 g(\tau) \, d\tau - \int_s^t T_0(t,\tau) P_0(\tau) g(\tau) \, d\tau$$

$$- \int_s^t T(t,\tau) Q_0(\tau) g(\tau) \, d\tau + \int_{-\infty}^t T_0(t,\tau) P_0(\tau) g(\tau) \, d\tau$$

$$- \int_t^{+\infty} T(t,\tau) Q_0(\tau) g(\tau) \, d\tau$$

$$= \int_s^t T_0(t,\tau) X_0 g(\tau) \, d\tau + \int_{-\infty}^s T_0(t,\tau) P_0(\tau) g(\tau) \, d\tau$$

$$- \int_s^{+\infty} T(t,\tau) Q_0(\tau) g(\tau) \, d\tau$$

$$= \int_s^t T_0(t,\tau) X_0 g(\tau) \, d\tau + T(t,s) v_s$$

and so identity (5.1) holds. In other words, v is a solution.

For the uniqueness of the former solution, it suffices to show that if $w_t = T(t,s) w_s$ for $t \ge s$ with $w \in \mathcal{E}$, then $w = 0$. Define

$$x_t = P(t) w_t \quad \text{and} \quad y_t = Q(t) w_t.$$

Then $w_t = x_t + y_t$ and by (3.2) we have

$$x_t = T(t, s)x_s \quad \text{and} \quad y_t = T(t, s)y_s$$

for $t \geq s$. Since $x_t = T(t, t - s)x_{t-s}$ for $s \geq 0$, we also have

$$\begin{aligned}
\|x_t\| &= \|T(t, t - s)x_{t-s}\| \\
&= \|T(t, t - s)P(t - s)w_{t-s}\| \\
&\leq Ne^{-\lambda s}\|w_{t-s}\| \leq Ne^{-\lambda s}|w|_{\mathcal{E}}.
\end{aligned}$$

Finally, letting $s \to +\infty$ we obtain $x_t = 0$ for $t \in \mathbb{R}$. Similarly, we have $y_t = 0$ for $t \in \mathbb{R}$ and so $w = 0$. $\qquad\square$

The converse of Theorem 5.1 also holds and is the principal result of this chapter. See Sec. 5.5 for the case of exponential dichotomies on $\mathbb{R}_0^+$.

Theorem 5.2 (Hyperbolicity via admissibility). *If the pair $(\mathcal{E}, \mathcal{F})$ is admissible for Eq. (2.12), then the equation has an exponential dichotomy on $\mathbb{R}$.*

The proof of Theorem 5.2 is given in Secs. 5.3–5.4. Putting together Theorems 5.1 and 5.2, we obtain a complete characterization of an exponential dichotomy on $\mathbb{R}$ via admissibility.

Theorem 5.3. *Equation (2.12) has an exponential dichotomy on $\mathbb{R}$ if and only if the pair $(\mathcal{E}, \mathcal{F})$ is admissible for the equation.*

At the end of Sec. 5.4 we consider the particular case of exponential contractions and we formulate a corresponding version of Theorem 5.2.

5.3   Invariant Subspaces and Projections

In this section we construct invariant subspaces for the dynamics defined by Eq. (2.12). These are obtained from the initial conditions that lead to bounded solutions, respectively, into the future and into the past.

Let R be the linear operator defined by $Rv = g$ in the domain $D(R)$ of all $v \in \mathcal{E}$ for which there exists a function $g \in \mathcal{F}$ such that (5.1) holds for $t, s \in \mathbb{R}$ with $t \geq s$. To show that R is well defined, take $g, h \in \mathcal{F}$ such that

$$v_t = T(t, s)v_s + \int_s^t T_0(t, \tau)X_0 g(\tau)\, d\tau$$

and

$$v_t = T(t, s)v_s + \int_s^t T_0(t, \tau)X_0 h(\tau)\, d\tau$$

for $t \geq s$. Since

$$\frac{1}{t-s} \int_s^t T_0(t,\tau) X_0 g(\tau)\, d\tau = \frac{1}{t-s} \int_s^t T_0(t,\tau) X_0 h(\tau)\, d\tau$$

and the maps

$$\tau \mapsto T_0(t,\tau) X_0 g(\tau) \quad \text{and} \quad \tau \mapsto T_0(t,\tau) X_0 h(\tau)$$

are locally integrable, taking the limit when $s \to t$ it follows from the Lebesgue differentiation theorem that $g(t) = h(t)$ for almost all $t \in \mathbb{R}$. This shows that R is well defined.

Lemma 5.1. *The operator $R \colon D(R) \to \mathcal{F}$ is closed.*

Proof. Consider a sequence $(v_m)_{m \in \mathbb{N}}$ in $D(R)$ converging to $v \in \mathcal{E}$ such that the sequence $Rv_m = g_m$ converges to $g \in \mathcal{F}$. Since

$$v_t^m = T(t,s) v_s^m + \int_s^t T_0(t,\tau) X_0 g_m(\tau)\, d\tau,$$

we obtain

$$
\begin{aligned}
v_t - T(t,s) v_s &= \lim_{m \to \infty} (v_t^m - T(t,s) v_s^m) \\
&= \lim_{m \to \infty} \int_s^t T_0(t,\tau) X_0 g_m(\tau)\, d\tau
\end{aligned}
\tag{5.2}
$$

for $t \geq s$. By Proposition 2.4, we have

$$
\begin{aligned}
\left\| \int_s^t T_0(t,\tau) X_0 (g_m(\tau) - g(\tau))\, d\tau \right\| &\leq \int_s^t K e^{\omega(t-\tau)} \| g_m(\tau) - g(\tau) \|\, d\tau \\
&\leq K e^{\omega(t-s)} (t-s+1) |g_m - g|_{\mathcal{F}}.
\end{aligned}
$$

Finally, since g_m converges to g in $\mathcal{F}$, we obtain

$$\lim_{m \to \infty} \int_s^t T_0(t,\tau) X_0 g_m(\tau)\, d\tau = \int_s^t T_0(t,\tau) X_0 g(\tau)\, d\tau$$

and together with (5.2) this implies that (5.1) holds for $t, s \in \mathbb{R}$ with $t \geq s$. Therefore, $Rv = g$ and $v \in D(R)$, showing that the operator R is closed. $\square$

Now observe that by the admissibility hypothesis in Theorem 5.2, the operator R has an inverse $S \colon \mathcal{F} \to \mathcal{E}$. By Lemma 5.1, it follows from the closed graph theorem that R is bounded and so its inverse S is also bounded.

Now we consider the subspaces composed of the initial conditions that lead to bounded solutions, respectively, into the future and into the past. Namely, for each $s \in \mathbb{R}$ let

$$E(s) = \left\{ \varphi \in C : \sup_{t \geq s} \| T(t,s) \varphi \| < +\infty \right\}$$

and let $F(s)$ be the set of all $\varphi \in C$ for which there exists a continuous function $v: (-\infty, s] \to \mathbb{R}^n$ with $v_s = \varphi$ such that $\sup_{t \le s}|v(t)| < +\infty$ and $v_t = T(t, \tau)v_\tau$ for $s \ge t \ge \tau$. Clearly, $E(s)$ and $F(s)$ are subspaces of C.

Lemma 5.2. *For each $s \in \mathbb{R}$ we have*

$$C = E(s) \oplus F(s). \tag{5.3}$$

Proof. Take $s \in \mathbb{R}$ and $\varphi \in C$. We define a function $\psi: [s-r, +\infty) \to \mathbb{R}^n$ by $\psi_s = \varphi$ and $\psi(t) = \varphi(0)$ for $t \ge s$. Moreover, define $g: \mathbb{R} \to \mathbb{R}^n$ by

$$g(t) = L(t)\psi_t = \int_{-r}^{0} d\eta(t, \theta)\psi(t + \theta)$$

for $t \ge s$ and $g(t) = 0$ for $t < s$. By (2.22) we have

$$|g|_{\mathcal{F}} = \sup_{t \in \mathbb{R}} \int_t^{t+1} |g(\tau)|\, d\tau \le \sup_{t \in \mathbb{R}} \int_t^{t+1} \Pi(\tau)\, d\tau \|\varphi\| < +\infty \tag{5.4}$$

and so $g \in \mathcal{F}$. Therefore, there exists $v \in D(R)$ such that $Rv = g$. Now let $u = v + \psi$. In a similar manner to that in the proof of Theorem 2.6, the function u is a solution of Eq. (2.12) with $u_s = v_s + \varphi$, that is,

$$u_t = T(t, s)(v_s + \varphi) \quad \text{for } t \ge s.$$

Note that

$$\sup_{t \ge s}|u(t)| \le \sup_{t \ge s}|v(t)| + \sup_{t \ge s}|\psi(t)| < +\infty$$

since $v \in \mathcal{E}$ and $|\psi(t)| \le \|\varphi\|$. Therefore, $v_s + \varphi \in E(s)$.

On the other hand, by (5.1) we have

$$v_t = T(t, \tau)v_\tau + \int_\tau^t T_0(t, \sigma)X_0 g(\sigma)\, d\sigma$$

and since $g(t) = 0$ for $t < s$, we conclude that $v_t = T(t, \tau)v_\tau$ for $s \ge t \ge \tau$. Therefore, $v_s \in F(s)$ and we obtain

$$\varphi = (v_s + \varphi) - v_s \in E(s) + F(s).$$

Finally, we show that

$$E(s) \cap F(s) = \{0\}.$$

Take $\varphi \in E(s) \cap F(s)$. There exists a continuous function $v: (-\infty, s] \to \mathbb{R}^n$ with $v_s = \varphi$ such that $\sup_{t \le s}|v(t)| < +\infty$ and $v_t = T(t, \tau)v_\tau$ for $s \ge t \ge \tau$. We also define a continuous function $u: \mathbb{R} \to \mathbb{R}^n$ by

$$u_t = \begin{cases} T(t, s)\varphi & \text{if } t \ge s, \\ v_t & \text{if } t \le s. \end{cases}$$

Clearly, $\sup_{t \in \mathbb{R}}|u(t)| < +\infty$ (because $\varphi \in E(s)$) and $u_t = T(t, \tau)u_\tau$ for $t \ge \tau$. Therefore, $Ru = 0$ and $u \in D(R)$. Since the operator R is invertible, we obtain $u = 0$ and so $\varphi = v_s = u_s = 0$. $\qquad\square$

From now on we denote by $P(s), Q(s): C \to C$ the projections associated with the decomposition in (5.3). The following result establishes condition 2 in Definition 3.2.

Lemma 5.3. *For each $t \geq s$, the linear operator*

$$T(t, s)|F(s): F(s) \to F(t) \tag{5.5}$$

is onto and invertible.

Proof. First assume that $T(t, s)\varphi = 0$ for some $\varphi \in F(s)$. Since $\varphi \in F(s)$, there exists a continuous function $v: (-\infty, s] \to \mathbb{R}^n$ with $v_s = \varphi$ such that $\sup_{\tau \leq s}|v(\tau)| < +\infty$ and $v_\tau = T(\tau, \bar{\tau})v_{\bar{\tau}}$ for $s \geq \tau \geq \bar{\tau}$. Define a continuous function $u: \mathbb{R} \to \mathbb{R}^n$ by

$$u_\tau = \begin{cases} T(\tau, s)\varphi & \text{if } \tau \geq s, \\ v_\tau & \text{if } \tau \leq s. \end{cases}$$

Clearly, $\sup_{\tau \in \mathbb{R}}|u(\tau)| < +\infty$ (because $u_\tau = 0$ for $\tau \geq t$) and $u_\tau = T(\tau, \bar{\tau})u_{\bar{\tau}}$ for $\tau \geq \bar{\tau}$. Hence, $Ru = 0$ and $u \in D(R)$. Since the operator R is invertible, we obtain $u = 0$ and so $\varphi = v_s = u_s = 0$. This shows that the map in (5.5) is injective.

Now take $\varphi \in F(t)$. There exists a continuous function $v: (-\infty, t] \to \mathbb{R}^n$ with $v_t = \varphi$ such that $\sup_{\tau \leq t}|v(\tau)| < +\infty$ and $v_\tau = T(\tau, \bar{\tau})v_{\bar{\tau}}$ for $t \geq \tau \geq \bar{\tau}$. Since

$$\varphi = v_t = T(t, s)v_s$$

and $v_s \in F(s)$, the map in (5.5) is also onto. $\qquad\square$

Now we establish condition 1 in Definition 3.2.

Lemma 5.4. *We have*

$$P(t)T(t, s) = T(t, s)P(s) \quad \text{for } t \geq s.$$

Proof. We first show that

$$T(t, s)E(s) \subset E(t) \quad \text{for } t \geq s.$$

Note that if $\varphi \in E(s)$, then $\sup_{\tau \geq s}\|T(\tau, s)\varphi\| < +\infty$ and so also

$$\sup_{\tau \geq t}\|T(\tau, t)T(t, s)\varphi\| \leq \sup_{\tau \geq s}\|T(\tau, s)\varphi\| < +\infty.$$

Therefore, $T(t, s)\varphi \in E(t)$. Now observe that by Lemma 5.3 we have

$$T(t, s)F(s) = F(t).$$

Since

$$T(t,s) = T(t,s)P(s) + T(t,s)Q(s),$$

it follows from the former discussion that

$$T(t,s)P(s)\varphi \in E(t) \quad \text{and} \quad T(t,s)Q(s)\varphi \in F(t)$$

for each $\varphi \in C$. Hence,

$$P(t)T(t,s)\varphi = T(t,s)P(s)\varphi \quad \text{and} \quad Q(t)T(t,s)\varphi = T(t,s)Q(s)\varphi,$$

which completes the proof of the lemma. $\qquad\square$

5.4 Existence of Exponential Behavior

To complete the proof of Theorem 5.2, it remains to establish the exponential bounds in (3.4). We first show that the projections $P(t)$ are uniformly bounded in t.

Lemma 5.5. *There exists $\widetilde{K} > 0$ such that*

$$\|P(s)\| \leq \widetilde{K} \quad \text{for } s \in \mathbb{R}. \tag{5.6}$$

Proof. Given $s \in \mathbb{R}$ and $\varphi \in C$, let ψ, g and v be as in the proof of Lemma 5.2. It follows from the proof that $P(s)\varphi = v_s + \varphi$. Denoting by S the inverse of R, we obtain

$$\|P(s)\varphi\| = \|v_s + \varphi\| \leq |v|_{\mathcal{E}} + \|\varphi\| = |Sg|_{\mathcal{E}} + \|\varphi\|.$$

On the other hand, by (5.4) we have

$$|Sg|_{\mathcal{E}} \leq \|S\| \cdot |g|_{\mathcal{F}} \leq \|S\|\gamma\|\varphi\|, \tag{5.7}$$

where $\gamma = \sup_{t \in \mathbb{R}} \int_t^{t+1} \Pi(\tau)\,d\tau$. Therefore,

$$\|P(s)\varphi\| \leq (\|S\|\gamma + 1)\|\varphi\|$$

which establishes (5.6) taking $\widetilde{K} = \|S\|\gamma + 1$. $\qquad\square$

Finally, we establish exponential bounds along the space $E(s)$ and $F(s)$. This turns out to be the most technical aspect of the proof of Theorem 5.2.

Lemma 5.6. *There exist constants $\lambda, D > 0$ such that*

$$\|T(t,s)P(s)\| \leq De^{-\lambda(t-s)} \quad \text{for } t \geq s.$$

Proof. Take $s \in \mathbb{R}$ and $\varphi \in E(s)$. Moreover, let ψ and g be as before in the proof of Lemma 5.2 and define $u_t = T(t, s)\varphi$ for $t \geq s$. Then the continuous function $v \colon \mathbb{R} \to \mathbb{R}^n$ given by

$$v(t) = \begin{cases} u(t) - \psi(t) & \text{if } t \geq s - r, \\ 0 & \text{if } t < s - r \end{cases}$$

is in $\mathcal{E}$ (because $\varphi \in E(s)$). For $t \geq s$ we have

$$v(t) = u(t) - \psi(t)$$

$$= \varphi(0) + \int_s^t L(\tau) u_\tau \, d\tau - \varphi(0)$$

$$= \int_s^t (v_\tau + \psi_\tau) \, d\tau$$

$$= \int_s^t \big(L(\tau) v_\tau + g(\tau) \big) \, d\tau.$$

Moreover, since $v(s) = \varphi(0) - \psi(s) = 0$, we obtain

$$v(t) = v(s) + \int_s^t \big(L(\tau) v_\tau + g(\tau) \big) \, d\tau.$$

In other words, v is a solution of the equation

$$v' = L(t) v_t + g(t)$$

on $[s - r, +\infty)$ and so also on $\mathbb{R}$ (recall that $v_t = 0$ and $g(t) = 0$ for $t < s$). This shows that $v \in R(D)$ and $Rv = g$. By (5.7), we obtain

$$\sup\{\|u_t\| : t \in [s, +\infty)\} \leq \sup\{\|v_t\| : t \in [s, +\infty)\} + \|\varphi\|$$
$$\leq |v|_{\mathcal{E}} + \|\varphi\| = |Sg|_{\mathcal{E}} + \|\varphi\|$$
$$\leq (\|S\|\gamma + 1)\|\varphi\|$$

and so

$$\|u_t\| \leq (\|S\|\gamma + 1)\|\varphi\| \quad \text{for } t \geq s. \tag{5.8}$$

The next step consists of showing that there exists $p \in \mathbb{N}$ (independent of s and φ) such that

$$\|u_t\| \leq \frac{1}{2}\|\varphi\| \quad \text{for } t - s \geq p. \tag{5.9}$$

Assume that there exists $t_0 > s$ such that $\|u_{t_0}\| > \|\varphi\|/2$ (otherwise there is nothing to show). By (5.8) and the same inequality with t, s and φ replaced, respectively, by t_0, τ and u_τ we obtain

$$\frac{1}{2(\|S\|\gamma + 1)}\|\varphi\| < \|u_\tau\| \leq (\|S\|\gamma + 1)\|\varphi\| \tag{5.10}$$

for $s \leq \tau \leq t_0$. Now let

$$w(t) = u(t) \int_{-\infty}^{t} \chi_{[s,t_0]}(\tau) \|u_\tau\|^{-1} \, d\tau$$

and

$$h(t) = \int_{-r}^{0} d\eta(t,\theta) \left(u(t+\theta) \int_{t+\theta}^{t} \chi_{[s,t_0]}(\tau) \|u_\tau\|^{-1} \, d\tau \right)$$
$$+ \chi_{[s,t_0]}(t) u(t) \|u_t\|^{-1}$$

for $t \in \mathbb{R}$. Note that w is continuous and since

$$|w(t)| \leq |u(t)| \int_{s}^{t_0} \|u_\tau\|^{-1} \, d\tau$$

$$\leq (\|S\|\gamma + 1)\|\varphi\| \int_{s}^{t_0} \|u_\tau\|^{-1} \, d\tau,$$

we have $w \in \mathcal{E}$. Now observe that

$$|h(t)| \leq \Pi(t) \sup_{\theta \in [-r,0]} \left| u(t+\theta) \int_{t+\theta}^{t} \|u_\tau\|^{-1} \, d\tau \right| + 1.$$

For $\tau \in [t+\theta, t]$ we have $t + \theta \in [\tau - r, \tau]$ and so

$$\|u_\tau\| \geq |u(t+\theta)|.$$

This implies that

$$|h(t)| \leq \Pi(t)r + 1, \tag{5.11}$$

and thus $h \in \mathcal{F}$. We also show that w is a solution of the equation

$$w' = L(t)w_t + h(t). \tag{5.12}$$

Writing $\alpha(t) = \chi_{[s,t_0]}(t)\|u_t\|^{-1}$, for $t \geq s$ we have

$$\int_{s}^{t} \left(L(\tau)w_\tau + h(\tau) \right) d\tau - \int_{s}^{t} \alpha(\tau)u(\tau) \, d\tau$$

$$= \int_{s}^{t} \int_{-r}^{0} d\eta(\tau,\theta) \left(w(\tau+\theta) + u(\tau+\theta) \int_{\tau+\theta}^{\tau} \alpha(\sigma) \, d\sigma \right) d\tau$$

$$= \int_{s}^{t} \int_{-r}^{0} d\eta(\tau,\theta) u(\tau+\theta) \int_{-\infty}^{\tau} \alpha(\sigma) \, d\sigma \, d\tau$$

$$= \int_{s}^{t} \left(L(\tau)u_\tau \right) \int_{-\infty}^{\tau} \alpha(\sigma) \, d\sigma \, d\tau$$

$$= \left(\int_{s}^{\tau} L(\sigma)u_\sigma \, d\sigma \int_{-\infty}^{\tau} \alpha(\sigma) \, d\sigma \right) \Bigg|_{\tau=s}^{\tau=t} - \int_{s}^{t} \int_{s}^{\tau} L(\sigma)u_\sigma \, d\sigma \, \alpha(\tau) \, d\tau$$

$$= \int_{s}^{t} L(\sigma)u_\sigma \, d\sigma \int_{-\infty}^{t} \alpha(\sigma) \, d\sigma - \int_{s}^{t} (u(\tau) - u(s))\alpha(\tau) \, d\tau$$

and so

$$\int_s^t \big(L(\tau)w_\tau + h(\tau)\big)\, d\tau = \int_s^t L(\sigma)u_\sigma\, d\sigma \int_{-\infty}^t \alpha(\sigma)\, d\sigma + \varphi(0) \int_s^t \alpha(\tau)\, d\tau$$

$$= \left(\varphi(0) + \int_s^t L(\sigma)u_\sigma\, d\sigma\right) \int_{-\infty}^t \alpha(\sigma)\, d\sigma$$

$$= u(t) \int_{-\infty}^t \alpha(\sigma)\, d\sigma = w(t).$$

Furthermore, $w_t = 0$ and $h(t) = 0$ for $t < s$. Thus, w is a solution of Eq. (5.12) and so $w \in D(R)$ and $Rw = h$. Moreover, by (5.11) we have

$$|w|_\mathcal{E} = |Sh|_\mathcal{E} \le \|S\| \cdot |h|_\mathcal{F} \le \|S\|(\gamma r + 1)$$

and using (5.10) we obtain

$$(\gamma r + 1)\|S\| \ge |w(t_0)| \ge |u(t_0)| \int_s^{t_0} \|u_\tau\|^{-1}\, d\tau$$

$$\ge \frac{\|u_{t_0}\|}{Ke^{\omega r}} \cdot \frac{t_0 - s}{(\|S\|\gamma + 1)\|\varphi\|}$$

$$\ge \frac{1}{2Ke^{\omega r}(\|S\|\gamma + 1)^2}(t_0 - s).$$

This shows that property (5.9) holds taking

$$p > 2Ke^{\omega r}(\gamma r + 1)\|S\|(\|S\|\gamma + 1)^2. \tag{5.13}$$

For $t \ge s$ write $t - s = kp + \tau$, with $k \in \mathbb{N}$ and $0 \le \tau < p$. By (5.8) and (5.9), we obtain

$$\|T(t, s)P(s)\varphi\| = \|T(s + kp + \tau, s)P(s)\varphi\|$$

$$\le \frac{1}{2^k}\|T(s + \tau, s)P(s)\varphi\|$$

$$\le \frac{\|S\|\gamma + 1}{2^k}\|P(s)\varphi\|$$

$$\le 2(\|S\|\gamma + 1)Me^{-(t-s)\log 2/p}\|\varphi\|$$

for $\varphi \in C$, which yields the desired statement. $\qquad\square$

Now we consider the space $F(s)$. We denote by $\overline{T}(s, t)$ the inverse of the linear operator in (5.5).

Lemma 5.7. *There exist constants $\lambda, D > 0$ such that*

$$\|\overline{T}(t, s)Q(s)\| \le De^{-\lambda(s-t)} \quad \text{for } t \le s.$$

Proof. Given $s \in \mathbb{R}$ and $\varphi \in F(s)$, let $\psi \colon (-\infty, s] \to \mathbb{R}^n$ be the continuous function defined by $\psi_s = \varphi$ and $\psi(t) = \varphi(-r)$ for $t \le s - r$. We also consider the function $g \colon \mathbb{R} \to \mathbb{R}^n$ given by

$$g(t) = L(t)\psi_t = \int_{-r}^{0} d\eta(t, \theta)\psi(t + \theta)$$

for $t \le s$ and $g(t) = 0$ for $t \ge s$. By (2.22) we have

$$|g|_{\mathcal{F}} = \sup_{t \in \mathbb{R}} \int_{t}^{t+1} |g(\tau)| \, d\tau \le \sup_{t \in \mathbb{R}} \int_{t}^{t+1} \Pi(\tau) \, d\tau \|\varphi\| < +\infty$$

and so $g \in \mathcal{F}$. By Lemma 5.3, one can define $u_t = \overline{T}(t, s)\varphi$ for $t \le s$. In a similar manner to that in the proof of Lemma 5.6, the continuous function $v \colon \mathbb{R} \to \mathbb{R}^n$ defined by

$$v(t) = \begin{cases} u(t) - \psi(t) & \text{if } t \le s, \\ 0 & \text{if } t > s \end{cases}$$

is in $D(R)$ and satisfies $Rv = g$. Therefore, by (5.7) we obtain

$$\begin{aligned} \sup\{\|u_t\| : t \in (-\infty, s]\} &\le \sup\{\|v_t\| : t \in (-\infty, s]\} + \|\varphi\| \\ &\le |v|_{\varepsilon} + \|\varphi\| = |Sg|_{\varepsilon} + \|\varphi\| \\ &\le (\|S\|\gamma + 1)\|\varphi\| \end{aligned}$$

and so

$$\|u_t\| \le (\|S\|\gamma + 1)\|\varphi\| \quad \text{for } t \le s. \tag{5.14}$$

We also show that there exists $p \in \mathbb{N}$ (independent of s and φ) such that

$$\|u_t\| \le \frac{1}{2}\|\varphi\| \quad \text{for } s - t \ge p. \tag{5.15}$$

Assume that there exists $t_0 < s$ such that $\|u_{t_0}\| > \|\varphi\|/2$. By (5.14) and the same inequality with t, s and φ replaced, respectively, by t_0, τ and u_τ we have

$$\frac{1}{2(\|S\|\gamma + 1)}\|\varphi\| < \|u_\tau\| \le (\|S\|\gamma + 1)\|\varphi\| \tag{5.16}$$

for $s \ge \tau \ge t_0$. Now let

$$w(t) = u(t) \int_{t}^{+\infty} \chi_{[t_0, s]}(\tau)\|u_\tau\|^{-1} \, d\tau$$

and

$$\begin{aligned} h(t) = &-\int_{-r}^{0} d\eta(t, \theta) \left(u(t + \theta) \int_{t+\theta}^{t} \chi_{[t_0, s]}(\tau)\|u_\tau\|^{-1} \, d\tau \right) \\ &- \chi_{[t_0, s]}(t)u(t)\|u_t\|^{-1} \end{aligned}$$

for $t \in \mathbb{R}$. In a similar manner to that in the proof of Lemma 5.6 we have $w \in \mathcal{E}$ and $h \in \mathcal{F}$ with $|h|_{\mathcal{F}} \leq \gamma r + 1$. We also have $w \in D(R)$ and $Rw = h$, which leads to

$$|w|_{\mathcal{E}} = |Sh|_{\mathcal{E}} \leq \|S\| \cdot |h|_{\mathcal{F}} \leq \|S\|(\gamma r + 1).$$

Finally, by (5.16) we obtain

$$
\begin{aligned}
(\gamma r + 1)\|S\| &\geq |w(t_0)| \geq |u(t_0)| \int_{t_0}^{s} \|u_\tau\|^{-1} \, d\tau \\
&\geq \frac{\|u_{t_0}\|}{K e^{\omega r}} \cdot \frac{s - t_0}{(\|S\|\gamma + 1)\|\varphi\|} \\
&\geq \frac{1}{2K e^{\omega r}(\|S\|\gamma + 1)^2}(s - t_0),
\end{aligned}
$$

which shows that property (5.15) holds taking p as in (5.13).

For $t \leq s$ write $s - t = kp + \tau$, with $k \in \mathbb{N}$ and $0 \leq \tau < p$. It follows from (5.14) and (5.15) that

$$
\begin{aligned}
\|\overline{T}(t, s)Q(s)\varphi\| &= \|\overline{T}(s - kp - \tau, s)Q(s)\varphi\| \\
&\leq \frac{1}{2^k}\|\overline{T}(s - \tau, s)Q(s)\varphi\| \\
&\leq \frac{\|S\|\gamma + 1}{2^k}\|Q(s)\varphi\| \\
&\leq 2(\|S\|\gamma + 1)M e^{-(s-t)\log 2/p}\|\varphi\|
\end{aligned}
$$

for $\varphi \in C$, which yields the desired statement. $\qquad\square$

It follows from the former lemmas that Eq. (2.12) has an exponential dichotomy on $\mathbb{R}$, thus completing the proof of Theorem 5.2.

We also consider briefly the particular case of exponential contractions. Given $s \in \mathbb{R}$ and $\varphi \in C$, let $g = g_{s,\varphi}$ be the function introduced in the proof of Lemma 5.2. Namely, let

$$
g_{s,\varphi}(t) = \begin{cases} L(t)\psi_t & \text{if } t \geq s, \\ 0 & \text{if } t < s, \end{cases}
$$

where $\psi \colon [s - r, +\infty) \to \mathbb{R}^n$ is defined by

$$
\psi(t) = \begin{cases} \varphi(t - s) & \text{if } t < s, \\ \varphi(0) & \text{if } t \geq s. \end{cases}
$$

By (5.4), we have $g_{s,\varphi} \in \mathcal{F}$. Using the function $g_{s,\varphi}$ one can formulate a version of Theorem 5.1 for exponential contractions.

Theorem 5.4. *If Eq. (2.12) has an exponential contraction on* $\mathbb{R}$, *then the pair* $(\mathcal{E}, \mathcal{F})$ *is admissible for the equation and given* $(s, \varphi) \in \mathbb{R} \times C$, *the unique function* $v \in \mathcal{E}$ *such that*

$$v_t = T(t, \tau)v_\tau + \int_\tau^t T_0(t, \tau)X_0 g_{s,\varphi}(\tau)\,d\tau \qquad (5.17)$$

for $t, \tau \in \mathbb{R}$ *with* $t \geq \tau$ *satisfies* $v_s = 0$.

Proof. The admissibility property is a particular case of Theorem 5.1, which allows one to apply Theorem 5.2. In particular, the additional property in Theorem 5.4 follows readily from Lemma 5.2 where we showed that $v_s \in F(s)$ for $g = g_{s,\varphi}$. Since Eq. (2.12) has an exponential contraction, we conclude that $v_s = 0$. $\qquad \square$

The converse of Theorem 5.4 also holds.

Theorem 5.5. *If the pair* $(\mathcal{E}, \mathcal{F})$ *is admissible for Eq. (2.12) and given* $(s, \varphi) \in \mathbb{R} \times C$, *the unique function* $v \in \mathcal{E}$ *such that (5.17) holds for* $t, \tau \in \mathbb{R}$ *with* $t \geq \tau$ *satisfies* $v_s = 0$, *then the equation has an exponential contraction on* $\mathbb{R}$.

Proof. By Theorem 5.2, Eq. (2.12) has an exponential dichotomy on $\mathbb{R}$. On the other hand, by Lemma 5.2, the unique function $v \in \mathcal{E}$ such that (5.17) holds satisfies $v_s \in F(s)$. But by hypothesis $v_s = 0$. Hence, the exponential dichotomy is in fact an exponential contraction. $\qquad \square$

5.5 One-Sided Admissibility

In this section we consider briefly the case of exponential dichotomies on the half-line and we give a corresponding characterization of the notion in terms of an admissibility property. Let $L(t) \colon C \to \mathbb{R}^n$, for $t \geq 0$, be bounded linear operators such that:

(1) the map $(t, v) \mapsto L(t)v$ satisfies the Carathéodory conditions on $\mathbb{R}_0^+ \times C$;
(2) the function Π in (2.11) satisfies a one-sided version of condition (2.22), that is,

$$\sup_{t \geq 0} \int_t^{t+1} \Pi(\tau)\,d\tau < +\infty. \qquad (5.18)$$

We first introduce appropriate Banach spaces. Let $\mathcal{E}^+$ be the set of all continuous functions $v\colon [-r, +\infty) \to \mathbb{R}^n$ such that

$$|v|_{\mathcal{E}^+} := \sup_{t \geq -r} |v(t)| < +\infty.$$

Note that $\mathcal{E}^+ = (\mathcal{E}^+, |\cdot|_{\mathcal{E}^+})$ is a Banach space. Given a closed subspace $B \subset C$, we denote by $\mathcal{E}_B^+$ the set of all functions $v \in \mathcal{E}^+$ with $v_0 \in B$. Clearly, $\mathcal{E}_B^+$ is a closed subspace of $\mathcal{E}^+$. Moreover, let $\mathcal{F}^+$ be the set of all measurable functions $g\colon [-r, +\infty) \to \mathbb{R}^n$ such that

$$|g|_{\mathcal{F}^+} := \sup_{t \geq -r} \int_t^{t+1} |g(\tau)|\, d\tau < +\infty,$$

identified if they are equal almost everywhere. Note that $\mathcal{F}^+ = (\mathcal{F}^+, |\cdot|_{\mathcal{F}^+})$ is a Banach space (see Proposition 5.1).

Theorem 5.6 (Admissibility via hyperbolicity). *If Eq. (2.12) has an exponential dichotomy on $\mathbb{R}_0^+$, then for each $g \in \mathcal{F}^+$ there exists a unique $v \in \mathcal{E}_B^+$ with $B = \operatorname{Im} Q(0)$ such that*

$$v_t = T(t,s)v_s + \int_s^t T_0(t,\tau)X_0 g(\tau)\, d\tau \tag{5.19}$$

for $t \geq s \geq 0$.

Proof. Given $g \in \mathcal{F}^+$, for each $t \geq 0$ let $v_t = x_t + y_t$, where

$$x_t = \int_0^t T_0(t,\tau)P_0(\tau)g(\tau)\, d\tau$$

and

$$y_t = -\int_t^{+\infty} T(t,\tau)Q_0(\tau)g(\tau)\, d\tau.$$

Repeating the proof of Theorem 5.1, we find that $v \in \mathcal{E}^+$ and that (5.19) holds for $t \geq s \geq 0$ (the arguments are exactly the same if we extend g to a function $g\colon \mathbb{R} \to \mathbb{R}^n$ letting $g(t) = 0$ for $t < 0$). Moreover, $P(0)v_0 = 0$ and so $v_0 \in B$. Therefore, $v \in \mathcal{E}_B^+$.

It remains to show that the solution is unique. As before, it suffices to verify that if $w_t = T(t,s)w_s$ for $t \geq s \geq 0$ with $w \in \mathcal{E}_B^+$, then $w = 0$. By Theorem 3.5 we have

$$\|Q(0)w_0\| = \|\overline{T}(0,t)Q(t)w_t\|$$
$$\leq Ne^{-\lambda t}\|w_t\|$$
$$\leq Ne^{-\lambda t}|w|_{\mathcal{E}^+}$$

for $t \geq 0$. Therefore, $w_0 = Q(0)w_0 = 0$ and so $w = 0$. $\qquad\square$

Now we establish the converse of Theorem 5.6.

Theorem 5.7 (Hyperbolicity via admissibility). *If there is a closed subspace $B \subset C$ such that for each $g \in \mathcal{F}^+$ there exists a unique $v \in \mathcal{E}_B^+$ such that (5.19) holds for $t \geq s \geq 0$, then Eq. (2.12) has an exponential dichotomy on $\mathbb{R}_0^+$.*

Proof. Let R_B be the linear operator defined by $R_B v = g$ in the domain $D(R_B)$ of all $v \in \mathcal{E}_B^+$ for which there exists $g \in \mathcal{F}^+$ such that (5.19) holds for $t \geq s \geq 0$. One can show as in Sec. 5.3 that R_B is well defined. It follows from the admissibility hypothesis that R_B has an inverse

$$S_B \colon \mathcal{F}^+ \to \mathcal{E}_B^+.$$

Moreover, the operator $R_B \colon D(R_B) \to \mathcal{F}^+$ is closed (see the proof of Lemma 5.1). By the closed graph theorem, R_B and S_B are bounded.

Now we consider subspaces composed of the initial conditions that lead to bounded solutions. Namely, for each $s \geq 0$ let

$$E(s) = \left\{ \varphi \in C : \sup_{t \geq s} \|T(t,s)\varphi\| < +\infty \right\} \quad \text{and} \quad F(s) = T(s,0)B. \quad (5.20)$$

Clearly, $E(s)$ and $F(s)$ are subspaces of C.

Lemma 5.8. *For each $s \geq 0$ we have*

$$C = E(s) \oplus F(s). \quad (5.21)$$

Proof of the lemma. Given $s \geq 0$ and $\varphi \in C$, let $\psi \colon [s - r, +\infty) \to \mathbb{R}^n$ be the function defined by $\psi_s = \varphi$ and $\psi(t) = \varphi(0)$ for $t \geq s$. Moreover, define $g \colon \mathbb{R}_0^+ \to \mathbb{R}^n$ by

$$g(t) = L(t)\psi_t = \int_{-r}^{0} d\eta(t,\theta)\psi(t + \theta)$$

for $t \geq s$ and $g(t) = 0$ for $t \in [0, s]$. As in (5.4), we have $g \in \mathcal{F}^+$ and so there exists $v \in \mathcal{E}_B^+$ such that $R_B v = g$. Moreover, as in the proof of Lemma 5.2, we have $v_s + \varphi \in E(s)$. On the other hand, by (5.19) we obtain

$$v_s = T(s,0)v_0$$

(recall that $g(t) = 0$ for $t \in [0, s]$). Since $v \in \mathcal{E}_B^+$, we have $v_s \in F(s)$ and so

$$\varphi = (v_s + \varphi) - v_s \in E(s) + F(s).$$

Given $\varphi \in E(s) \cap F(s)$, take $\psi \in B$ such that $\varphi = T(s,0)\psi$. We define a continuous map $u \colon [-r, +\infty) \to \mathbb{R}^n$ by $u_t = T(t,0)\psi$ for $t \geq 0$. Note that $\sup_{t \geq -r} |u(t)| < +\infty$ and $u_0 \in B$. Moreover,

$$u_t = T(t,\tau)u_\tau \quad \text{for } t \geq \tau \geq 0.$$

This shows that $u \in D(R_B)$ and $R_B u = 0$. Since R_B is invertible, we obtain $u = 0$ and so $\varphi = u_s = 0$. $\qquad\square$

We denote by $P(s), Q(s) \colon C \to C$ the projections associated with the decomposition in (5.21).

Lemma 5.9. *For each $t \geq s \geq 0$, the map*

$$T(t,s)|F(s) \colon F(s) \to F(t) \tag{5.22}$$

is onto and invertible.

Proof of the lemma. Take $\varphi \in F(s)$ such that $T(t,s)\varphi = 0$. Then there exists $\psi \in B$ such that $\varphi = T(s,0)\psi$ and we define a continuous map $v \colon [-r, +\infty) \to \mathbb{R}^n$ by

$$v_\tau = T(\tau, 0)\psi.$$

Note that $\sup_{\tau \geq -r} |v(\tau)| < +\infty$ and $v_\tau = T(\tau, \overline{\tau})v_{\overline{\tau}}$ for $\tau \geq \overline{\tau} \geq 0$. Therefore, $v \in D(R_B)$ and $R_B v = 0$. Since R_B is invertible, we have $v = 0$ and so $\varphi = v_s = 0$. In other words, the map in (5.22) is injective.

In order to show that the map in (5.22) is onto, take $\varphi \in F(t)$ and $\psi \in B$ such that $\varphi = T(t,0)\psi$. Note that $T(s,0)\psi \in F(s)$ and

$$\varphi = T(t,s)T(s,0)\psi.$$

This completes the proof of the lemma. $\qquad\qquad\qquad\qquad\qquad\square$

Repeating the proofs, respectively, of Lemmas 5.4 and 5.5 we obtain

$$P(t)T(t,s) = T(t,s)P(s) \quad \text{for } t \geq s \geq 0.$$

Moreover, there exists $\widetilde{K} > 0$ such that

$$\|P(s)\| \leq \widetilde{K} \quad \text{for } s \geq 0.$$

In fact one can take

$$\widetilde{K} = \|S_B\| \sup_{t \geq 0} \int_t^{t+1} \Pi(\tau)\, d\tau + 1.$$

Now let $\overline{T}(s,t)$ be the inverse of the map in (5.22). Repeating the proofs of Lemmas 5.6 and 5.7 we find that there exist constants $\lambda, D > 0$ such that

$$\|T(t,s)P(s)\| \leq De^{-\lambda(t-s)} \quad \text{and} \quad \|\overline{T}(s,t)Q(t)\| \leq De^{-\lambda(t-s)}$$

for $t \geq s \geq 0$. It follows from the former results that Eq. (2.12) has an exponential dichotomy on $\mathbb{R}_0^+$. $\qquad\qquad\qquad\qquad\qquad\square$

Putting together Theorems 5.6 and 5.7, we obtain a complete characterization of an exponential dichotomy on $\mathbb{R}_0^+$ via admissibility.

Theorem 5.8. *Equation (2.12) has an exponential dichotomy on $\mathbb{R}_0^+$ if and only if there exists a closed subspace $B \subset C$ such that for each $g \in \mathcal{F}^+$ there exists a unique $v \in \mathcal{E}_B^+$ such that (5.19) holds for $t \geq s \geq 0$.*

One can also formulate appropriate versions of Theorems 5.6 and 5.7 for exponential contractions. The following result is a simple consequence of Theorem 5.6.

Theorem 5.9. *If Eq. (2.12) has an exponential contraction on* $\mathbb{R}$*, then for each* $g \in \mathcal{F}^+$ *there exists a unique* $v \in \mathcal{E}^+$ *with* $v_0 = 0$ *such that (5.19) holds for* $t \geq s \geq 0$.

The converse of Theorem 5.9 also holds.

Theorem 5.10. *If there exists a unique* $v \in \mathcal{E}^+$ *with* $v_0 = 0$ *such that (5.19) holds for* $t \geq s \geq 0$*, then Eq. (2.12) has an exponential contraction on* $\mathbb{R}_0^+$.

Proof. By Theorem 5.7, Eq. (2.12) has an exponential dichotomy on $\mathbb{R}_0^+$, taking the closed space $B = \{0\}$. Moreover, it follows from (5.20) that

$$F(t) = T(t,0)B = \{0\}$$

is an unstable space for the exponential dichotomy. Hence, Eq. (2.12) has in fact an exponential contraction on $\mathbb{R}_0^+$. $\square$

Chapter 6

Robustness and Parameters

In this chapter we use the characterizations of an exponential dichotomy and of an exponential contraction in terms of an admissibility property, given in Chap. 5, to establish the robustness of the notions under sufficiently small perturbations. We also consider perturbations depending on a parameter and we obtain a parameterized version of the robustness property both for Lipschitz and smooth perturbations. In particular, we give a different proof of the two-sided robustness result in Chap. 4.

6.1 Robustness via Admissibility

In this section we use the characterization of an exponential dichotomy on $\mathbb{R}_0^+$ in terms of an admissibility property, given by Theorems 5.6 and 5.7, to establish the robustness of the notion under sufficiently small perturbations. In a certain sense this is a simpler approach than the one given in Chap. 4, although at the expense of being less explicit in giving the projections for the exponential dichotomy of the new equation. Sometimes it is quite helpful to have a more explicit representation and so it really depends on the particular problem at hand which approach is more convenient.

We consider linear perturbations of Eq. (2.12) of the form (4.1), that is,

$$v' = (L(t) + M(t))v_t,$$

where $L(t), M(t) \colon C \to \mathbb{R}^n$, for $t \geq 0$, are bounded linear operators such that:

(1) the functions $(t, v) \mapsto L(t)v$ and $(t, v) \mapsto M(t)v$ satisfy the Carathéodory conditions on $\mathbb{R}_0^+ \times C$;
(2) the functions Π in (2.11) and $t \mapsto \|M(t)\|$ are locally integrable.

In addition, we assume that condition (5.18) holds.

Theorem 6.1.   *Assume that Eq. (2.12) has an exponential dichotomy on $\mathbb{R}_0^+$ and that there exists $\delta > 0$ such that*

$$\sup_{t \geq 0} \int_t^{t+1} \|M(\tau)\| \, d\tau \leq \delta. \tag{6.1}$$

For any sufficiently small δ, Eq. (4.1) has an exponential dichotomy on $\mathbb{R}_0^+$.

Proof. Let $\mathcal{E}^+$, $\mathcal{E}_B^+$ and $\mathcal{F}^+$ be the Banach spaces introduced in Sec. 5.5. Since Eq. (2.12) has an exponential dichotomy on $\mathbb{R}_0^+$, by Theorem 5.6 there exists a closed subspace $B \subset C$ such that the operator $R_B \colon D(R_B) \to \mathcal{F}^+$ introduced in the proof of Theorem 5.7 is invertible. For $v \in D(R_B)$ we consider the graph norm

$$\|v\|_{R_B} = |v|_{\mathcal{E}^+} + |R_B v|_{\mathcal{F}^+}.$$

Since the operator R_B is closed, $(D(R_B), \|\cdot\|_{R_B})$ is a Banach space. Moreover, the operator

$$R_B \colon (D(R_B), \|\cdot\|_{R_B}) \to \mathcal{F}^+ \tag{6.2}$$

is bounded and from now on we denote it simply by R_B.

Now let $\widetilde{T}(t,s)$ be the evolution family associated with Eq. (4.1). Moreover, let $\widetilde{R}_B$ be the linear operator defined by $\widetilde{R}_B v = g$ in the domain $D(\widetilde{R}_B)$ of all $v \in \mathcal{E}_B^+$ for which there exists $g \in \mathcal{F}^+$ such that

$$v_t = \widetilde{T}(t,s)v_s + \int_s^t \widetilde{T}_0(t,\tau) X_0 g(\tau) \, d\tau$$

for $t \geq s \geq 0$. One can show as in Section 5.3 that $\widetilde{R}_B$ is well defined. We define a linear operator $G \colon \mathcal{E}_B^+ \to \mathcal{F}^+$ by

$$(Gv)(t) = M(t)v_t \quad \text{for } t \geq 0.$$

It follows from (6.1) that

$$\int_t^{t+1} |(Gv)(\tau)| \, d\tau \leq \int_t^{t+1} \|M(\tau)\| \, d\tau |v|_{\mathcal{E}^+} \leq \delta |v|_{\mathcal{E}^+} \tag{6.3}$$

for $v \in \mathcal{E}_B^+$. Hence, G is a well defined bounded linear operator.

Take $v \in \mathcal{E}_B^+$ and $g \in \mathcal{F}^+$ such that $\widetilde{R}_B v = g$. Moreover, let $T(t,s)$ be the evolution family associated with Eq. (2.12). By (4.3) we have

$$v_t = \widetilde{T}(t,s)v_s + \int_s^t \widetilde{T}_0(t,\tau)X_0 g(\tau)\,d\tau$$

$$= T(t,s)v_s + \int_s^t T_0(t,\tau)X_0 M(\tau)\widetilde{T}(\tau,s)v_s\,d\tau$$

$$+ \int_s^t T_0(t,\tau)X_0 g(\tau)\,d\tau + \int_s^t \int_\tau^t T_0(t,r)X_0 M(r)\widetilde{T}_0(r,\tau)X_0 g(\tau)\,dr\,d\tau$$

$$= T(t,s)v_s + \int_s^t T_0(t,r)X_0 M(r)\widetilde{T}(r,s)v_s\,dr$$

$$+ \int_s^t T_0(t,\tau)X_0 g(\tau)\,d\tau + \int_s^t \int_s^r T_0(t,r)X_0 M(r)\widetilde{T}_0(r,\tau)X_0 g(\tau)\,d\tau\,dr$$

$$= T(t,s)v_s + \int_s^t T_0(t,\tau)X_0 g(\tau)\,d\tau$$

$$+ \int_s^t T_0(t,r)X_0 M(r)\left(\widetilde{T}(r,s)v_s + \int_s^r \widetilde{T}_0(r,\tau)X_0 g(\tau)\,d\tau\right)dr$$

$$= T(t,s)v_s + \int_s^t T_0(t,r)X_0\big(g(r) + M(r)v_r\big)\,dr$$

$$\tag{6.4}$$

for $t \geq s \geq 0$. Therefore, $v \in D(R_B)$ and

$$R_B v = \widetilde{R}_B v + G v.$$

In particular, we have $D(\widetilde{R}_B) \subset D(R_B)$. Reversing the arguments, we also obtain $D(R_B) \subset D(\widetilde{R}_B)$ and so

$$D(\widetilde{R}_B) = D(R_B) \quad \text{and} \quad R_B = \widetilde{R}_B + G. \tag{6.5}$$

It follows from (6.3) that

$$|(\widetilde{R}_B - R_B)v|_{\mathcal{F}^+} \leq \delta|v|_{\mathcal{E}^+} \leq \delta\|v\|_{R_B} \tag{6.6}$$

for $v \in D(R_B)$. Hence, the linear operator

$$\widetilde{R}_B \colon (D(R_B), \|\cdot\|_{R_B}) \to \mathcal{F}^+ \tag{6.7}$$

is bounded. Moreover, it follows from (6.6) and the invertibility of R_B that if δ is sufficiently small, then the operator $\widetilde{R}_B$ is also invertible. Hence, applying Theorem 5.7 we conclude that Eq. (4.1) has an exponential dichotomy on $\mathbb{R}_0^+$. $\qquad\square$

Now we consider the particular case of exponential contractions.

Theorem 6.2. *Assume that Eq. (2.12) has an exponential contraction on $\mathbb{R}_0^+$ and that there exists $\delta > 0$ such that (6.1) holds. For any sufficiently small δ, Eq. (4.1) has an exponential contraction on $\mathbb{R}_0^+$.*

Proof. By Theorem 5.9, one can repeat the proof of Theorem 6.1 taking $B = \{0\}$ to show that the operator $\widetilde{R}_B$ in (6.7) is invertible. Since $B = \{0\}$, it follows from Theorem 5.10 that Eq. (2.12) has an exponential contraction on $\mathbb{R}_0^+$. $\qquad\square$

6.2   Lipschitz Parameter Dependence

In this section we consider perturbations depending on a parameter and we obtain a parameterized version of the robustness result in Theorem 6.1. More precisely, we consider perturbations that are Lipschitz on the parameter. See Sec. 6.3 for the case of smooth perturbations.

Let Λ be a Banach space (the parameter space). We consider the perturbed equation

$$v' = (L(t) + M(t, \rho))v_t, \qquad (6.8)$$

where $L(t), M(t, \rho) \colon C \to \mathbb{R}^n$, for $t \geq 0$ and $\rho \in \Lambda$, are bounded linear operators such that the maps

$$(t, v) \mapsto L(t)v \quad \text{and} \quad (t, v, \rho) \mapsto M(t, \rho)v \qquad (6.9)$$

are continuous, respectively, on $\mathbb{R}_0^+ \times C$ and on $\mathbb{R}_0^+ \times C \times \Lambda$.

Proposition 6.1. *Assume that Eq. (2.12) has an exponential dichotomy on $\mathbb{R}_0^+$ and that there exists $\delta > 0$ such that*

$$\sup_{t \geq 0} \int_t^{t+1} \|M(\tau, \rho)\| \, d\tau \leq \delta \quad \text{for } \rho \in \Lambda. \qquad (6.10)$$

For any sufficiently small δ, Eq. (6.8) has an exponential dichotomy on $\mathbb{R}_0^+$ for each $\rho \in \Lambda$.

Proof. It follows from Theorem 6.1 that if δ is sufficiently small, then Eq. (6.8) has an exponential dichotomy on $\mathbb{R}_0^+$ for each $\rho \in \Lambda$. $\qquad\square$

Now we consider perturbations that are Lipschitz on the parameter and we obtain a corresponding parameterized version of the robustness result in Theorem 6.1.

Theorem 6.3. *Under the assumptions of Proposition 6.1, if δ is sufficiently small and there exists $d > 0$ such that*

$$\sup_{t \geq 0} \int_t^{t+1} \|M(\tau, \rho) - M(\tau, \varsigma)\| \, d\tau \leq d\|\rho - \varsigma\| \tag{6.11}$$

for $\rho, \varsigma \in \Lambda$, then for the projections $P_\rho(t)$ onto the stable spaces there exists $c > 0$ (independent of t) such that the map $\rho \mapsto P_\rho(t)$ is Lipschitz on each ball of radius c, with a Lipschitz constant independent of t and the ball.

Proof. By Proposition 6.1, if δ is sufficiently small, then Eq. (6.8) has an exponential dichotomy on $\mathbb{R}_0^+$ for each $\rho \in \Lambda$.

We continue to consider the operator R_B in (6.2). For each $\rho \in \Lambda$ let $R_{B,\rho}$ be the linear operator defined by $R_{B,\rho} v = g$ in the domain $D(R_{B,\rho})$ of all $v \in \mathcal{E}_B^+$ for which there exists $g \in \mathcal{F}^+$ such that

$$v_t = T_\rho(t, s) v_s + \int_s^t T_\rho(t, \tau) X_0 g(\tau) \, d\tau$$

for $t \geq s \geq 0$, where $T_\rho(t, s)$ is the evolution family associated with Eq. (6.8). One can show as in Section 5.3 that $R_{B,\rho}$ is well defined. Moreover, for each $\rho \in \Lambda$ we define a linear operator $G_\rho \colon \mathcal{E}_B^+ \to \mathcal{F}^+$ by

$$(G_\rho v)(t) = M(t, \rho) v_t$$

for $t \geq 0$. It follows from (6.10) that

$$\int_t^{t+1} |(G_\rho v)(\tau)| \, d\tau \leq \int_t^{t+1} \|M(\tau, \rho)\| \, d\tau |v|_{\mathcal{E}^+} \leq \delta |v|_{\mathcal{E}^+} \tag{6.12}$$

for $v \in \mathcal{E}_B^+$. Hence, G_ρ is a well defined bounded linear operator. By (6.5), for $\rho \in \Lambda$ we have

$$D(R_{B,\rho}) = D(R_B) \quad \text{and} \quad R_B = R_{B,\rho} + G_\rho. \tag{6.13}$$

It follows from (6.12) and (6.13) that

$$|(R_{B,\rho} - R_B) v|_{\mathcal{F}^+} \leq \delta |v|_{\mathcal{E}^+} \leq \delta \|v\|_{R_B} \tag{6.14}$$

for $v \in D(R_B)$ and $\rho \in \Lambda$. Hence, the linear operator

$$R_{B,\rho} \colon (D(R_B), \|\cdot\|_{R_B}) \to \mathcal{F}^+$$

is bounded for each $\rho \in \Lambda$. Moreover, it follows from (6.14) and the invertibility of R_B that if δ is sufficiently small, then the operator $R_{B,\rho}$ is invertible for $\rho \in \Lambda$.

Let $S_{B,\rho}\colon \mathcal{F}^+ \to \mathcal{E}_B^+$ be the inverse of $R_{B,\rho}$. It follows from the proof of Lemma 5.8 that the projections $P_\rho(s)$ onto the stable spaces are given for each $\varphi \in C$ by

$$P_\rho(s)\varphi = (S_{B,\rho}g_{\rho,s})_s + \varphi, \tag{6.15}$$

where

$$g_{\rho,s}(t) = \begin{cases} (L(t) + M(t,\rho))\psi_t & \text{if } t \geq s, \\ 0 & \text{if } t \in [0,s], \end{cases} \tag{6.16}$$

taking $\psi\colon [s-r,+\infty) \to \mathbb{R}^n$ such that $\psi_s = \varphi$ and $\psi(t) = \varphi(0)$ for $t \geq s$. By (6.15), we have

$$\begin{aligned}
\|P_\rho(s)\varphi - P_\varsigma(s)\varphi\| &= \|(S_{B,\rho}g_{\rho,s} - S_{B,\varsigma}g_{\varsigma,s})_s\| \\
&\leq |S_{B,\rho}g_{\rho,s} - S_{B,\varsigma}g_{\varsigma,s}|_{\mathcal{E}^+} \\
&\leq \|S_{B,\rho}g_{\rho,s} - S_{B,\varsigma}g_{\varsigma,s}\|_{R_B} \\
&\leq \|S_{B,\rho}\| \cdot |g_{\rho,s} - g_{\varsigma,s}|_{\mathcal{F}^+} + \|S_{B,\rho} - S_{B,\varsigma}\| \cdot |g_{\varsigma,s}|_{\mathcal{F}^+}.
\end{aligned}$$

Note that

$$\begin{aligned}
|g_{\varsigma,s}(t)| &\leq \big(\|L(t)\| + \|M(\rho,t)\|\big)\|\psi_t\| \\
&\leq \big(\Pi(t) + \|M(\rho,t)\|\big)\|\varphi\|
\end{aligned}$$

and so $|g_{\varsigma,s}|_{\mathcal{F}^+} \leq \gamma\|\varphi\|$, where

$$\gamma = \sup_{t \geq 0} \int_t^{t+1} \Pi(\tau)\,d\tau + \delta$$

(using (6.10)). Moreover, since

$$g_{\rho,s}(t) - g_{\varsigma,s}(t) = (M(t,\rho) - M(t,\varsigma))\psi_t,$$

it follows from (6.11) that

$$|g_{\rho,s} - g_{\varsigma,s}|_{\mathcal{F}^+} \leq d\|\rho - \varsigma\| \cdot \|\varphi\|.$$

Therefore, for each $\varphi \in C$ we have

$$\|P_\rho(s)\varphi - P_\varsigma(s)\varphi\| \leq d\|S_{B,\rho}\| \cdot \|\rho - \varsigma\| \cdot \|\varphi\| + \gamma\|S_{B,\rho} - S_{B,\varsigma}\| \cdot \|\varphi\|$$

and so

$$\|P_\rho(s) - P_\varsigma(s)\| \leq d\|S_{B,\rho}\| \cdot \|\rho\| + \gamma\|S_{B,\rho} - S_{B,\varsigma}\|. \tag{6.17}$$

Now we show that the map $\rho \mapsto R_{B,\rho}$ is Lipschitz. Indeed, it follows from (6.11) and (6.13) that

$$\begin{aligned}
|(R_{B,\rho} - R_{B,\varsigma})v|_{\mathcal{F}^+} &= |(G_\rho - G_\varsigma)v|_{\mathcal{F}^+} \\
&= \sup_{t \geq 0} \int_t^{t+1} |(M(\tau,\rho) - M(\tau,\varsigma))v_\tau|\,d\tau \\
&\leq d|\rho - \varsigma\| \cdot |v|_{\mathcal{E}^+} \\
&\leq d\|\rho - \varsigma\| \cdot \|v\|_{R_B}
\end{aligned}$$

for $v \in D(R_B)$ and $\rho, \varsigma \in \Lambda$. Hence,

$$\|R_{B,\rho} - R_{B,\varsigma}\| \le d\|\rho - \varsigma\| \tag{6.18}$$

for $\rho, \varsigma \in \Lambda$. On the other hand, given two linear operators A and B, we have the identity

$$B^{-1} - A^{-1} = A^{-1} \sum_{m=1}^{\infty} \left[(A - B)A^{-1} \right]^m$$

(see for example [75]) and so

$$\|B^{-1} - A^{-1}\| \le \|A^{-1}\| \sum_{m=1}^{\infty} \|(A - B)A^{-1}\|^m$$
$$\le \frac{\|A^{-1}\|^2 \|A - B\|}{1 - \|A^{-1}\| \cdot \|A - B\|}.$$

Thus, given $\rho \in \Lambda$, if r is sufficiently small and $\|\rho - \varsigma\| < r$, it follows from (6.18) that

$$\|S_{B,\rho} - S_{B,\varsigma}\| \le \frac{\|S_{B,\rho}\|^2 \|R_{B,\rho} - R_{B,\varsigma}\|}{1 - \|S_{B,\rho}\| \cdot \|R_{B,\rho} - R_{B,\varsigma}\|}$$
$$\le \frac{d\|S_{B,\rho}\|^2 \|\rho - \varsigma\|}{1 - rd\|S_{B,\rho}\|}.$$

Hence, by (6.17) we have

$$\|P_\rho(s) - P_\varsigma(s)\| \le \left(d + \gamma \frac{d\|S_{B,\rho}\|}{1 - rd\|S_{B,\rho}\|} \right) \|S_{B,\rho}\| \cdot \|\rho - \varsigma\|. \tag{6.19}$$

Finally, we estimate $\|S_{B,\rho}\|$. Note that

$$S_{B,\rho} = (R_B - G_\rho)^{-1} = \left[R_B(\mathrm{Id} - S_B G_\rho) \right]^{-1}$$
$$= (\mathrm{Id} - S_B G_\rho)^{-1} S_B = \sum_{m=0}^{\infty} (S_B G_\rho)^m S_B,$$

where $S_B = R_B^{-1}$, and so

$$\|S_{B,\rho}\| \le \sum_{m=0}^{\infty} \|S_B G_\rho\|^m \|S_B\| \le \frac{\|S_B\|}{1 - \delta\|S_B\|}.$$

Hence, by (6.19), there exist $K', c > 0$ (independent of ρ and s) such that

$$\|P_\rho(s) - P_\varsigma(s)\| \le K'\|\rho - \varsigma\|$$

whenever $\|\rho - \varsigma\| < c$ and δ is sufficiently small. This completes the proof of the theorem. $\qquad\square$

6.3   Smooth Parameter Dependence

In this section we establish a smooth version of the robustness result in Theorem 6.3. Instead of the hypothesis in (6.11), we assume that the map $\rho \mapsto M(t, \rho)v$ is of class C^1 for each $t \geq 0$ and $v \in C$. We denote by ∂ the partial derivative with respect to ρ.

Theorem 6.4 (Smooth parameter dependence). *Under the assumptions of Proposition 6.1, if δ is sufficiently small and for each $\rho \in \Lambda$ and $\varepsilon > 0$ there exists $c > 0$ such that*

$$\sup_{t \geq 0} \int_t^{t+1} \|\partial M(\tau, \rho) - \partial M(\tau, \varsigma)\|\, d\tau \leq \varepsilon \tag{6.20}$$

whenever $\|\rho - \varsigma\| \leq c$, then for the projections $P_\rho(t)$ onto the stable spaces the map $\rho \mapsto P_\rho(t)$ is of class C^1 for each $t \geq 0$.

Proof. By Proposition 6.1, for each $\rho \in \Lambda$ Eq. (6.8) has an exponential dichotomy on $\mathbb{R}_0^+$. It remains to establish the smooth dependence of the projections on the parameter. We use the same notation as in the proof of Theorem 6.3.

It follows from (6.13) and (6.20) that $\rho \mapsto R_{B,\rho}$ is of class C^1. Hence, the map $\rho \mapsto S_{B,\rho}$ is also C^1. Now we define a map $H_{\rho,s} \colon \Lambda \to \mathcal{F}^+$ by

$$(H_{\rho,s}\varsigma)(t) = (\partial M(t, \rho)\varsigma)\psi_t,$$

with ψ_t as in (6.16) (note that ψ depends on s). It follows again from (6.20) that the map

$$\rho \mapsto H_{\rho,s} \in \mathcal{L}(\Lambda, \mathcal{F}^+),$$

where $\mathcal{L}(\Lambda, \mathcal{F}^+)$ is the set of all bounded linear operators from Λ into $\mathcal{F}^+$, is of class C^1. Writing $v_t = C_t v$, we obtain

$$\frac{P_{\rho+\varepsilon\varsigma}(s)\varphi - P_\rho(s)\varphi}{\varepsilon} = \frac{C_s(S_{B,\rho+\varepsilon\varsigma}g_{\rho+\varepsilon\varsigma,s} - S_{B,\rho}g_{\rho,s})}{\varepsilon}$$

$$= C_s\left(S_{B,\rho+\varepsilon\varsigma}\frac{g_{\rho+\varepsilon\varsigma,s} - g_{\rho,s}}{\varepsilon} + \frac{S_{B,\rho+\varepsilon\varsigma} - S_{B,\rho}}{\varepsilon}g_{\rho,s}\right)$$

$$\to C_s\left(S_{B,\rho}H_{\rho,s}\varsigma + \left(\frac{d}{d\rho}S_{B,\rho}\varsigma\right)g_{\rho,s}\right)$$

when $\varepsilon \to 0$, because $\rho \mapsto g_{\rho,s}$ and $\rho \mapsto S_{B,\rho}$ are C^1 functions. Therefore,

$$\frac{d}{d\rho}P_\rho(s)\varphi = C_s\left(S_{B,\rho}H_{\rho,s} + \frac{d}{d\rho}S_{B,\rho}g_{\rho,s}\right).$$

Since all functions on the right-hand side are continuous on ρ, we conclude that $\rho \mapsto (d/d\rho)P_\rho(s)$ is also continuous. This completes the proof of the theorem. $\qquad\square$

6.4 Two-Sided Robustness

This section is a two-sided version of the former sections. Namely, we use the characterization of an exponential dichotomy on $\mathbb{R}$ in terms of an admissibility property, given by Theorems 5.1 and 5.2, to provide a new proof of the robustness property in Theorem 4.1. The argument does not give the stable and unstable spaces of the perturbed equation so explicitly, but it is considerably shorter (after having the characterization of an exponential dichotomy in terms of an admissibility property). We also consider the case of perturbations depending on a parameter.

Let $L(t), M(t) \colon C \to \mathbb{R}^n$, for $t \in \mathbb{R}$, be bounded linear operators such that:

(1) the functions $(t, v) \mapsto L(t)v$ and $(t, v) \mapsto M(t)v$ satisfy the Carathéodory conditions on $\mathbb{R} \times C$;
(2) the functions Π in (2.11) and $t \mapsto \|M(t)\|$ are locally integrable.

In addition, we assume that condition (2.22) holds.

Theorem 6.5. *Assume that Eq. (2.12) has an exponential dichotomy on $\mathbb{R}$ and that there exists $\delta > 0$ such that (4.4) holds. For any sufficiently small δ, Eq. (4.1) has an exponential dichotomy on $\mathbb{R}$.*

Proof. Let $\mathcal{E}$ and $\mathcal{F}$ be the Banach spaces introduced in Sec. 5.1 and consider the evolution family $\widetilde{T}(t, s)$ associated with Eq. (4.1). Moreover, let $\widetilde{R}$ be the linear operator defined by $\widetilde{R}v = g$ in the domain $D(\widetilde{R})$ of all $v \in \mathcal{E}$ for which there exists $g \in \mathcal{F}$ such that

$$v_t = \widetilde{T}(t, s)v_s + \int_s^t \widetilde{T}_0(t, \tau)X_0 g(\tau) \, d\tau$$

for $t, s \in \mathbb{R}$ with $t \geq s$. One can show as in Sec. 5.3 that $\widetilde{R}$ is well defined. Now take $v \in \mathcal{E}$ and $g \in \mathcal{F}$ such that $\widetilde{R}v = g$. Proceeding as in (6.4) we obtain

$$v_t = T(t, s)v_s + \int_s^t T_0(t, r)X_0 \big(g(r) + M(r)v_r\big) \, dr$$

for $t \geq s$. We also define a linear operator $G \colon \mathcal{E} \to \mathcal{F}$ by $(Gv)(t) = M(t)v_t$. Note that

$$|(Gv)(t)| = |M(t)v_t| \leq \|M(t)\| \cdot |v|_{\mathcal{E}}$$

for $t \in \mathbb{R}$. Hence, it follows from (4.4) that

$$\int_t^{t+1} |(Gv)(\tau)| \, d\tau \leq \delta |v|_{\mathcal{E}} \tag{6.21}$$

for $t \in \mathbb{R}$ and so G is a well defined bounded linear operator. Furthermore, it follows from (6.4) that
$$D(R) = D(\widetilde{R}) \quad \text{and} \quad R = \widetilde{R} + G.$$
For $v \in D(R)$ we consider the graph norm
$$\|v\|_R = |v|_{\mathcal{E}} + |Rv|_{\mathcal{F}}.$$
By Lemma 5.1, R is closed and so $(D(R), \|\cdot\|_R)$ is a Banach space. Moreover, the operator
$$R \colon (D(R), \|\cdot\|_R) \to \mathcal{F}$$
is bounded and from now on we denote it simply by R. By (6.21) we have
$$|(R - \widetilde{R})v|_{\mathcal{F}} \leq \delta |v|_{\mathcal{E}} \leq \delta \|v\|_R \qquad (6.22)$$
for $v \in D(R)$. By Theorem 5.1, the operator R is invertible. Hence, it follows from (6.22) that if δ is sufficiently small, then $\widetilde{R}$ is also invertible. Applying Theorem 5.2, we conclude that Eq. (4.1) has an exponential dichotomy on $\mathbb{R}$. □

One can also consider perturbations depending on a parameter and obtain corresponding two-sided versions of Theorems 6.3 and 6.4. Namely, let Λ be a Banach space (the parameter space). We consider the perturbed Eq. (6.8), where $L(t), M(t, \rho) \colon C \to \mathbb{R}^n$, for $t \in \mathbb{R}$ and $\rho \in \Lambda$, are bounded linear operators such that the maps in (6.9) are continuous, respectively, on $\mathbb{R} \times C$ and on $\mathbb{R} \times C \times \Lambda$.

Proposition 6.2. *Assume that Eq. (2.12) has an exponential dichotomy on $\mathbb{R}$ and that there exists $\delta > 0$ such that*
$$\sup_{t \in \mathbb{R}} \int_t^{t+1} \|M(\tau, \rho)\| \, d\tau \leq \delta \quad \text{for } \rho \in \Lambda.$$
For any sufficiently small δ, Eq. (6.8) has an exponential dichotomy on $\mathbb{R}$ for each $\rho \in \Lambda$.

Proof. It follows from Theorem 6.5 that if δ is sufficiently small, then Eq. (6.8) has an exponential dichotomy on $\mathbb{R}$ for each $\rho \in \Lambda$. □

Now we consider perturbations that are Lipschitz on the parameter and we formulate a two-sided version of Theorem 6.3. The proof is analogous and so we omit it.

Theorem 6.6. *Under the assumptions of Proposition 6.2, if δ is sufficiently small and there exists $d > 0$ such that*
$$\sup_{t \in \mathbb{R}} \int_t^{t+1} \|M(\tau, \rho) - M(\tau, \varsigma)\| \, d\tau \leq d \|\rho - \varsigma\|$$

for $\rho, \varsigma \in \Lambda$, then for the projections $P_\rho(t)$ onto the stable spaces there exists $c > 0$ (independent of t) such that the map $\rho \mapsto P_\rho(t)$ is Lipschitz on each ball of radius c, with a Lipschitz constant independent of t and the ball.

We also formulate a two-sided version of Theorem 6.4. Again the proof is analogous and so we omit it.

Theorem 6.7. *Under the assumptions of Proposition 6.2, if δ is sufficiently small and for each $\rho \in \Lambda$ and $\varepsilon > 0$ there exists $c > 0$ such that*

$$\sup_{t \in \mathbb{R}} \int_t^{t+1} \|\partial M(\tau, \rho) - \partial M(\tau, \varsigma)\| \, d\tau \leq \varepsilon$$

whenever $\|\rho - \varsigma\| \leq c$, then for the projections $P_\rho(t)$ onto the stable spaces the map $\rho \mapsto P_\rho(t)$ is of class C^1 for each $t \in \mathbb{R}$.

PART III
Nonlinear Stability

Chapter 7

Lipschitz Perturbations

This chapter is a first step towards the discussion of the behavior of the solutions of a linear delay equation under sufficiently small nonlinear perturbations (in the sense that the Lipschitz constant is sufficiently small). We refer to Chaps. 8 and 9 for the study of smooth perturbations. As a prototype, we start with a discussion of the nonlinear perturbations of an exponential contraction. We also consider the more general case of perturbations depending on a parameter and we describe how the solutions depend on this parameter. We then construct Lipschitz stable and unstable invariant manifolds for any sufficiently small Lipschitz perturbation of an exponential dichotomy. More precisely, we show that the set of initial conditions leading to a bounded forward global solution is a graph of a Lipschitz function. A similar statement holds for the bounded backward global solutions, leading to the construction of the unstable manifold, although in this case one also needs to show that the solutions exist and are unique, besides being global, which does not follow from the general results in Chap. 2. In addition, we establish a partial version of the Grobman–Hartman theorem in the theory of ordinary differential equations.

7.1 Perturbations of Exponential Contractions

In this section we establish the persistence of the asymptotic stability of an exponential contraction on an interval $I \subset \mathbb{R}$ containing $\mathbb{R}_0^+$ under sufficiently small nonlinear perturbations. We also consider perturbations depending on a parameter and we study how the solutions depend on this parameter. The general case of perturbations of exponential dichotomies is considered in Secs. 7.2 and 7.4 (see Chap. 8 for the construction of smooth invariant manifolds).

More precisely, we consider the perturbations

$$v' = L(t)v_t + g(t, v_t) \tag{7.1}$$

of Eq. (2.12), where:

(1) $L(t)\colon C \to \mathbb{R}^n$, for $t \in I$, are bounded linear operators such that the
    map $(t, v) \mapsto L(t)v$ is continuous on $I \times C$ and condition (2.22) holds;
(2) $g\colon I \times C \to \mathbb{R}^n$ is continuous and $g(t, 0) = 0$ for all $t \in I$.

We shall also assume that g is Lipschitz in the second variable. Given
$(s, \varphi) \in I \times C$, we denote by $v(\cdot, s, \varphi)$ the unique solution of Eq. (7.1) on
its maximal interval $[s - r, a)$ with $v_s = \varphi$. Notice that $v_t(\cdot, s, 0) = 0$ since
$g(t, 0) = 0$ for all t.

Theorem 7.1. *Assume that Eq. (2.12) has an exponential contraction on
an interval I containing $\mathbb{R}_0^+$ and that there exists $\delta > 0$ such that*

$$|g(t, u) - g(t, v)| \le \delta \|u - v\| \tag{7.2}$$

*for every $t \in I$ and $u, v \in C$. Then each solution $v(\cdot, s, \varphi)$ of Eq. (7.1) is
defined on $[s - r, +\infty)$ and*

$$\|v_t(\cdot, s, \varphi) - v_t(\cdot, s, \psi)\| \le Ne^{-(\lambda - \delta N)(t-s)}\|\varphi - \psi\| \tag{7.3}$$

for every $t, s \in I$ with $t \ge s$ and $\varphi, \psi \in C$.

Proof. We first show that each solution $v(\cdot, s, \varphi)$ is forward global, that is,
is defined on the interval $[s - r, +\infty)$. By the variation of constants formula
(see Theorem 2.7), we have

$$v_t = T(t, s)\varphi + \int_s^t T_0(t, \tau)X_0 g(\tau, v_\tau)\, d\tau$$

for $t \in [s, a)$. Since Eq. (2.12) has an exponential contraction, we obtain

$$\|v_t\| \le Ne^{-\lambda(t-s)}\|\varphi\| + \int_s^t \|T_0(t, \tau)\| \cdot |g(\tau, v_\tau)|\, d\tau$$

$$\le Ne^{-\lambda(t-s)}\|\varphi\| + \delta N \int_s^t e^{-\lambda(t-\tau)}\|v_\tau\|\, d\tau$$

or, equivalently,

$$e^{\lambda t}\|v_t\| \le Ne^{\lambda s}\|\varphi\| + \delta N \int_s^t e^{\lambda \tau}\|v_\tau\|\, d\tau.$$

Applying Gronwall's lemma yields the inequality

$$e^{\lambda t}\|v_t\| \le Ne^{\lambda s}\|\varphi\|e^{\delta N(t-s)},$$

and so

$$\|v_t\| \le Ne^{-(\lambda-\delta N)(t-s)}\|\varphi\| \le N\|\varphi\|$$

for $t \in [s, a)$. Therefore,

$$|L(t)v_t + g(t, v_t)| \le \Pi(t)N\|\varphi\| + \delta N\|\varphi\|$$

for $t \in [s, a)$. Since Π is locally integrable, one can proceed as in the proof of Theorem 2.4 to conclude that v can be continued to the interval $[s-r, +\infty)$.

Now take $s \in I$ and $\varphi, \psi \in C$. Moreover, let $v_t = v_t(\cdot, s, \varphi)$ and $w_t = v_t(\cdot, s, \psi)$. Again by the variation of constants formula, we have

$$v_t = T(t, s)\varphi + \int_s^t T_0(t, \tau)X_0 g(\tau, v_\tau)\, d\tau$$

and

$$w_t = T(t, s)\psi + \int_s^t T_0(t, \tau)X_0 g(\tau, w_\tau)\, d\tau$$

for $t \ge s$. Hence,

$$v_t - w_t = T(t, s)(\varphi - \psi) + \int_s^t T_0(t, \tau)X_0(g(\tau, v_\tau) - g(\tau, w_\tau))\, d\tau.$$

Since Eq. (2.12) has an exponential contraction, we obtain

$$\|v_t - w_t\| \le Ne^{-\lambda(t-s)}\|\varphi - \psi\| + \int_s^t \|T_0(t, \tau)\| \cdot |g(\tau, v_\tau) - g(\tau, w_\tau)|\, d\tau$$

$$\le Ne^{-\lambda(t-s)}\|\varphi - \psi\| + \delta N \int_s^t e^{-\lambda(t-\tau)}\|v_\tau - w_\tau\|\, d\tau$$

$$\le Ne^{-\lambda(t-s)}\|\varphi - \psi\| + \delta Ne^{-\lambda(t-s)} \int_s^t e^{\lambda(\tau-s)}\|v_\tau - w_\tau\|\, d\tau.$$

Letting $\alpha(t) = e^{\lambda(t-s)}\|v_t - w_t\|$ we have

$$\alpha(t) \le N\|\varphi - \psi\| + \delta N \int_s^t \alpha(\tau)\, d\tau$$

for $t \ge s$. Finally, applying Gronwall's lemma we obtain

$$\alpha(t) \le Ne^{\delta N(t-s)}\|\varphi - \psi\|$$

for $t \ge s$, which establishes property (7.3). $\qquad\square$

Notice that taking $\delta < \lambda/N$ in Theorem 7.1 yields an exponential decay in (7.3). Moreover, taking $\psi = 0$ in (7.3) we obtain

$$\|v_t(\cdot, s, \varphi)\| \le Ne^{-(\lambda - \delta N)(t-s)}\|\varphi\| \qquad (7.4)$$

for every $t, s \in I$ with $t \ge s$ and $\varphi \in C$.

Now we consider the more general case of perturbations of Eq. (2.12) depending on a parameter and we study how the solutions depend on this parameter. Namely, we consider nonlinear perturbations of this equation of the form

$$v' = L(t)v_t + g(t, v_t, \rho), \qquad (7.5)$$

for some continuous function $g: I \times C \times \Lambda \to \mathbb{R}^n$, where $\Lambda = (\Lambda, |\cdot|)$ is a Banach space (the parameter space). We assume that

$$g(t, 0, \rho) = 0 \quad \text{for } t \in I, \rho \in \Lambda \qquad (7.6)$$

and that g is Lipschitz in the second variable uniformly on the parameter ρ. Given $(s, \varphi, \rho) \in I \times C \times \Lambda$, we denote by $v(\cdot, s, \varphi, \rho)$ the unique solution of Eq. (7.5) on its maximal interval with $v_s = \varphi$. Notice that $v_t(\cdot, s, 0, \rho) = 0$.

Theorem 7.2. *Assume that Eq. (2.12) has an exponential contraction on an interval I containing $\mathbb{R}_0^+$ and that there exists $\delta > 0$ such that*

$$|g(t, u, \rho) - g(t, v, \rho)| \le \delta\|u - v\| \qquad (7.7)$$

and

$$|g(t, v, \rho) - g(t, v, \varsigma)| \le \delta|\rho - \varsigma| \cdot \|v\| \qquad (7.8)$$

for every $t \in I$, $u, v \in C$ and $\rho, \varsigma \in \Lambda$. Then

$$\|v_t(\cdot, s, \varphi, \rho) - v_t(\cdot, s, \psi, \rho)\| \le Ne^{-(\lambda - \delta N)(t-s)}\|\varphi - \psi\| \qquad (7.9)$$

and

$$\|v_t(\cdot, s, \varphi, \rho) - v_t(\cdot, s, \varphi, \varsigma)\| \le Ne^{-(\lambda - 2\delta N)(t-s)}|\rho - \varsigma| \cdot \|\varphi\| \qquad (7.10)$$

for every $t, s \in I$ with $t \ge s$, $\varphi, \psi \in C$ and $\rho, \varsigma \in \Lambda$.

Proof. In view of condition (7.7), it follows from Theorem 7.1 that each solution $v(\cdot, s, \varphi, \rho)$ is forward global. Moreover, property (7.9) follows readily from property (7.3). Now we establish property (7.10). Write

$$v_t = v_t(\cdot, s, \varphi, \rho) \quad \text{and} \quad w_t = v_t(\cdot, s, \varphi, \varsigma).$$

By (7.4), (7.7) and (7.8), we have

$$
\begin{aligned}
|g(\tau, v_\tau, \rho) - g(\tau, w_\tau, \varsigma)| &\leq |g(\tau, v_\tau, \rho) - g(\tau, v_\tau, \varsigma)| \\
&\quad + |g(\tau, v_\tau, \varsigma) - g(\tau, w_\tau, \varsigma)| \\
&\leq \delta |\rho - \varsigma| \cdot \|v_\tau\| + \delta \|v_\tau - w_\tau\| \\
&\leq \delta N e^{-(\lambda - \delta N)(\tau - s)} |\rho - \varsigma| \cdot \|\varphi\| + \delta \|v_\tau - w_\tau\|.
\end{aligned}
$$

Since Eq. (2.12) has an exponential contraction, we obtain

$$
\begin{aligned}
\|v_t - w_t\| &\leq \int_s^t \|T_0(t, \tau)\| \cdot |g(\tau, v_\tau, \rho) - g(\tau, w_\tau, \varsigma)| \, d\tau \\
&\leq \delta N^2 \int_s^t e^{-\lambda(t-\tau)} e^{-(\lambda - \delta N)(\tau - s)} |\rho - \varsigma| \cdot \|\varphi\| \, d\tau \\
&\quad + \delta N \int_s^t e^{-\lambda(t-\tau)} \|v_\tau - w_\tau\| \, d\tau \\
&\leq \delta N^2 e^{-\lambda(t-s)} |\rho - \varsigma| \cdot \|\varphi\| \int_s^t e^{\delta N(\tau - s)} \, d\tau \\
&\quad + \delta N e^{-\lambda(t-s)} \int_s^t e^{\lambda(\tau - s)} \|v_\tau - w_\tau\| \, d\tau \\
&\leq N e^{-\lambda(t-s)} e^{\delta N(t-s)} |\rho - \varsigma| \cdot \|\varphi\| \\
&\quad + \delta N e^{-\lambda(t-s)} \int_s^t e^{\lambda(\tau - s)} \|v_\tau - w_\tau\| \, d\tau.
\end{aligned}
$$

Hence, letting

$$
\alpha(t) = e^{\lambda(t-s)} \|v_t - w_t\|,
$$

for each $\bar{t} > s$ we have

$$
\alpha(t) \leq N e^{\delta N(\bar{t} - s)} |\rho - \varsigma| \cdot \|\varphi\| + \delta N \int_s^t \alpha(\tau) \, d\tau
$$

for $t \in [s, \bar{t}]$. Applying Gronwall's lemma yields the inequality

$$
\alpha(t) \leq N e^{\delta N(\bar{t} - s) + \delta N(t - s)} |\rho - \varsigma| \cdot \|\varphi\|
$$

for $t \in [s, \bar{t}]$. In particular, taking $t = \bar{t}$ we obtain

$$
\alpha(\bar{t}) \leq N e^{2\delta N(\bar{t} - s)} |\rho - \varsigma| \cdot \|\varphi\|
$$

and since $\bar{t}$ is arbitrary, this establishes property (7.10). $\qquad \square$

7.2 Lipschitz Stable Invariant Manifolds

In this section we construct a Lipschitz stable invariant manifold for
Eq. (7.1) when the linear Eq. (2.12) has an exponential dichotomy on an
interval $I \subset \mathbb{R}$ containing $\mathbb{R}_0^+$ and g is a sufficiently small Lipschitz perturbation. More precisely, we show that for each $s \in I$, the set of initial
conditions (s, φ) leading to a forward bounded solution of Eq. (7.1) is a
graph of a Lipschitz function over the stable space $E(s)$. We continue to
assume that conditions 1 and 2 in Sec. 7.1 hold.

We first introduce the notion of stable set. As in Sec. 7.1, we assume
that the perturbation g in (7.1) is Lipschitz in the second variable. Given
$(s, \varphi) \in I \times C$, we denote by $v(\cdot, s, \varphi)$ the unique solution of Eq. (7.1) on
its maximal interval with $v_s = \varphi$.

Definition 7.1. The *stable set* V^s of Eq. (7.1) is the set of all initial conditions $(s, \varphi) \in I \times C$ for which $v(\cdot, s, \varphi)$ is defined and bounded on the
interval $[s - r, +\infty)$.

The stable set has the following invariance property.

Proposition 7.1. *If* $(s, \varphi) \in V^s$, *then* $(t, v_t(\cdot, s, \varphi)) \in V^s$ *for all* $t \geq s$.

Proof. It suffices to show that

$$v(t, \sigma, v_\sigma(\cdot, s, \varphi)) = v(t, s, \varphi) \tag{7.11}$$

for all $t \geq \sigma \geq s$. Indeed, since $v(\cdot, s, \varphi)$ is bounded, this implies that
$v(\cdot, \sigma, v_\sigma(\cdot, s, \varphi))$ is also bounded and so $(\sigma, v_\sigma(\cdot, s, \varphi)) \in V^s$ for all $\sigma \geq s$.

In order to establish (7.11), first notice that by the variation of constants
formula we have

$$v_\sigma = T(\sigma, s)\varphi + \int_s^\sigma T_0(\sigma, \tau) X_0 g(\tau, v_\tau)\, d\tau$$

for $\sigma \geq s$. Hence,

$$T(t, \sigma)v_\sigma = T(t, \sigma)\left(T(\sigma, s)\varphi + \int_s^\sigma T_0(\sigma, \tau) X_0 g(\tau, v_\tau)\, d\tau \right)$$

$$= T(t, s)\varphi + \int_s^\sigma T_0(t, \tau) X_0 g(\tau, v_\tau)\, d\tau$$

and so

$$T(t, \sigma)v_\sigma + \int_\sigma^t T_0(t, \tau) X_0 g(\tau, v_\tau)\, d\tau$$

$$= T(t, s)\varphi + \int_s^t T_0(t, \tau) X_0 g(\tau, v_\tau)\, d\tau = v_t$$

for $t \geq \sigma$. This establishes identity (7.11). $\square$

The following result gives an alternative characterization of the stable set when Eq. (2.12) has an exponential dichotomy.

Proposition 7.2. *Assume that Eq. (2.12) has an exponential dichotomy on an interval I containing $\mathbb{R}_0^+$. Then its stable set is composed of the pairs $(s, \varphi) \in I \times C$ for which there exists a bounded continuous function $v \colon [s - r, +\infty) \to \mathbb{R}^n$ such that $v_s = \varphi$ and*

$$
\begin{aligned}
v_t = T(t, s)P(s)\varphi &+ \int_s^t T_0(t, \tau)P_0(\tau)g(\tau, v_\tau)\, d\tau \\
&- \int_t^{+\infty} \overline{T}(t, \tau)Q_0(\tau)g(\tau, v_\tau)\, d\tau
\end{aligned}
\tag{7.12}
$$

for every $t \geq s$.

Proof. First assume that v is a bounded solution of Eq. (7.1) on $[s-r, +\infty)$ with $v_s = \varphi$. By Theorem 3.6, we have

$$
\begin{aligned}
P(\tau)v_\tau &= T(\tau, \sigma)P(\sigma)v_\sigma + \int_\sigma^\tau T_0(\tau, u)P_0(u)g(u, v_u)\, du, \\
Q(\tau)v_\tau &= T(\tau, \sigma)Q(\sigma)v_\sigma + \int_\sigma^\tau \overline{T}(\tau, u)Q_0(u)g(u, v_u)\, du
\end{aligned}
\tag{7.13}
$$

for $\tau \geq \sigma \geq s$. Taking $\sigma = t$, we write the second identity in (7.13) in the form

$$
Q(t)v_t = \overline{T}(t, \tau)Q(\tau)v_\tau - \int_t^\tau \overline{T}(t, u)Q_0(u)g(u, v_u)\, du.
$$

By the second inequality in (3.23), since v is bounded we obtain

$$
\|\overline{T}(t, \tau)Q(\tau)v_\tau\| \leq Ne^{-\lambda(\tau - t)} \sup_{u \geq s} \|v_u\| \to 0
$$

when $\tau \to +\infty$. Hence,

$$
Q(t)v_t = - \int_t^{+\infty} \overline{T}(t, u)Q_0(u)g(u, v_u)\, du,
$$

which added to the first identity in (7.13) with $\tau = t$ and $\sigma = s$ yields (7.12).

Now we assume that property (7.12) holds with $v_s = \varphi$. Taking $t = s$, we obtain

$$
v_s = P(s)v_s - \int_s^{+\infty} \overline{T}(s, u)Q_0(u)g(u, v_u)\, du
$$

and hence,

$$
Q(s)v_s = - \int_s^{+\infty} \overline{T}(s, u)Q_0(u)g(u, v_u)\, du.
$$

Therefore,

$$T(t,s)v_s + \int_s^t T_0(t,u)X_0 g(u, v_u)\, du$$

$$= T(t,s)P(s)v_s + \int_s^t T_0(t,u)P_0(u)g(u, v_u)\, du$$

$$+ \int_s^t T(t,u)Q_0(u)g(u, v_u)\, du + T(t,s)Q(s)v_s$$

$$= v_t + \int_t^{+\infty} \overline{T}(t,u)Q_0(u)g(u, v_u)\, du$$

$$+ \int_s^t T(t,u)Q_0(u)g(u, v_u)\, du$$

$$- T(t,s)\int_s^{+\infty} \overline{T}(s,u)Q_0(u)g(u, v_u)\, du = v_t$$

and so v is a solution of Eq. (7.1).   □

Using the characterization of the stable set in Proposition 7.2, one can show that when Eq. (2.12) has an exponential dichotomy, all bounded solutions of Eq. (7.1) decay exponentially.

Proposition 7.3 (Exponential decay). *Assume that Eq. (2.12) has an exponential dichotomy on an interval I containing $\mathbb{R}_0^+$ and that there exists $\delta > 0$ such that (7.2) holds for every $t \in I$ and $u, v \in C$. Then, for any sufficiently small δ, each bounded solution $v\colon [s-r, +\infty) \to \mathbb{R}^n$ of Eq. (7.1) with $v_s = \varphi$ satisfies*

$$\|v_t\| \le 2N e^{-(\lambda - 2\delta N)(t-s)}\|P(s)\varphi\| \tag{7.14}$$

for every $t \ge s$.

Proof. By Proposition 7.2, each bounded solution $v\colon [s - r, +\infty) \to \mathbb{R}^n$ of Eq. (7.1) with $v_s = \varphi$ satisfies (7.12). On the other hand, by Theorem 3.5 and (7.2) together with the fact that $g(t, 0) = 0$, it follows from (7.12) that

$$\|v_t\| \le N e^{-\lambda(t-s)}\|P(s)\varphi\| + \delta N \int_s^t e^{-\lambda(t-\tau)}\|v_\tau\|\, d\tau$$

$$+ \delta N \int_t^{+\infty} e^{\lambda(t-\tau)}\|v_\tau\|\, d\tau. \tag{7.15}$$

Before proceeding with the proof of the proposition, we establish an auxiliary result that will be used various times in the chapter

Lemma 7.1. *Given $s \in I$, let $x \colon [s, +\infty) \to \mathbb{R}_0^+$ be a bounded continuous function such that*

$$x(t) \leq \beta e^{-\gamma(t-s)} + \delta D \int_s^t e^{-\gamma(t-\tau)} x(\tau)\, d\tau + \delta D \int_t^{+\infty} e^{\gamma(t-\tau)} x(\tau)\, d\tau \quad (7.16)$$

for every $t \geq s$. Then, for any sufficiently small δ, we have

$$x(t) \leq 2\beta e^{-(\gamma-2\delta D)(t-s)} \quad \text{for } t \geq s. \tag{7.17}$$

Proof of the lemma. We first show that $x(t) \leq y(t)$ for $t \geq s$, where y is the unique bounded continuous function satisfying the integral equation

$$y(t) = \beta e^{-\gamma(t-s)} + \delta D \int_s^t e^{-\gamma(t-\tau)} y(\tau)\, d\tau + \delta D \int_t^{+\infty} e^{\gamma(t-\tau)} y(\tau)\, d\tau \quad (7.18)$$

for $t \geq s$. One can easily verify that y is a solution of the ordinary differential equation

$$y'' - \gamma^2 \left(1 - \frac{2\delta D}{\gamma}\right) y = 0. \tag{7.19}$$

In order that y is bounded for $t \geq s$, we must have

$$y(t) = y(s) e^{-\overline{\gamma}(t-s)}, \quad \text{where } \overline{\gamma} = \gamma \sqrt{1 - \frac{2\delta D}{\gamma}}.$$

Substituting $y(t)$ in (7.18) and taking $t = s$, for any sufficiently small δ we obtain

$$y(s) = \beta + \delta D y(s) \int_s^{+\infty} e^{(\gamma+\overline{\gamma})(s-\tau)}\, d\tau = \beta + y(s)\frac{\delta D}{\gamma + \overline{\gamma}}.$$

This yields

$$y(s) = \frac{\beta}{1 - \delta D/(\gamma + \overline{\gamma})}$$

and thus,

$$y(t) = \frac{\beta}{1 - \delta D/(\gamma + \overline{\gamma})} e^{-\overline{\gamma}(t-s)}.$$

Now let

$$z(t) = x(t) - y(t) \quad \text{for } t \geq s.$$

It follows from (7.16) and (7.18) that

$$z(t) \le \delta D \int_s^t e^{-\gamma(t-\tau)} z(\tau) \, d\tau + \delta D \int_t^{+\infty} e^{\gamma(t-\tau)} z(\tau) \, d\tau \qquad (7.20)$$

for $t \ge s$. Since the functions x and y are bounded, $z = \sup_{t \ge s} z(t)$ is finite and taking the supremum in (7.20) we obtain

$$z \le \delta D z \sup_{t \ge s} \int_s^t e^{-\gamma(t-\tau)} \, d\tau + \delta D z \sup_{t \ge s} \int_t^{+\infty} e^{\gamma(t-\tau)} \, d\tau + \frac{2\delta D}{\gamma} z.$$

Finally, taking δ sufficiently small, we conclude that $z \le 0$ and so $z(t) \le 0$ for all $t \ge s$. This yields the upper bound

$$x(t) \le y(t) = \frac{\beta}{1 - \delta D/(\gamma + \overline{\gamma})} e^{-\overline{\gamma}(t-s)} \qquad (7.21)$$

for $t \ge s$. Note that

$$\overline{\gamma} = \sqrt{\gamma(\gamma - 2\delta D)} \ge \gamma - 2\delta D$$

and so for δ sufficiently small we have

$$\frac{\beta}{1 - \delta D/(\gamma + \overline{\gamma})} \le 2\beta.$$

Hence, it follows from (7.21) that property (7.17) holds. $\qquad \square$

In view of (7.15), applying Lemma 7.1 taking

$$x(t) = \|v_t\|, \quad \beta = N\|P(s)\varphi\|, \quad \gamma = \lambda \quad \text{and} \quad D = N,$$

we obtain inequality (7.14). $\qquad \square$

Now we formulate the Lipschitz stable manifold theorem. Assume that Eq. (2.12) has an exponential dichotomy on an interval I and denote the corresponding stable and unstable spaces, respectively, by $E(s)$ and $F(s)$. Let $\mathcal{S}^L$ be the set of all continuous functions

$$z \colon \big\{(s,a) \in I \times C : a \in E(s)\big\} \to C$$

such that for each $s \in I$:

(1) $z(s,0) = 0$ and $z(s, E(s)) \subset F(s)$;
(2) for $a, \overline{a} \in E(s)$ we have

$$\|z(s,a) - z(s,\overline{a})\| \le \|a - \overline{a}\|. \qquad (7.22)$$

For each function $z \in \mathcal{S}^L$ we consider its graph

$$\operatorname{graph} z = \big\{(s, a + z(s,a)) : (s,a) \in I \times E(s)\big\} \subset I \times C. \qquad (7.23)$$

Theorem 7.3 (Lipschitz stable manifold). *If Eq. (2.12) has an exponential dichotomy on an interval I containing $\mathbb{R}_0^+$ and there exists $\delta > 0$ such that (7.2) holds for every $t \in I$ and $u, v \in C$, then, for any sufficiently small δ, there exists a function $z \in \mathcal{S}^L$ such that $V^s = \text{graph } z$. Moreover,*

$$\|w_t^a - w_t^{\bar{a}}\| \leq 2Ne^{-(\lambda - 2\delta N)(t-s)}\|a - \bar{a}\| \tag{7.24}$$

for every $t, s \in I$ with $t \geq s$ and $a, \bar{a} \in E(s)$, where $w_t^a = v_t(\cdot, s, a + z(s,a))$.

Proof. Take $s \in I$ and $a \in E(s)$. We consider the set $\mathcal{X}^a$ of all continuous functions $v \colon [s - r, +\infty) \to \mathbb{R}^n$ with $P(s)v_s = a$ such that

$$\|v_t\| \leq 3Ne^{-(\lambda - 2\delta N)(t-s)}\|a\|$$

for all $t \geq s$. One can easily verify that $\mathcal{X}^a$ is a complete metric space when equipped with the norm

$$|v|_{\mathcal{X}^a} := \sup\left\{\|v_t\|e^{(\lambda - 2\delta N)(t-s)} : t \geq s\right\}.$$

We define an operator J on $\mathcal{X}^a$ by

$$(Jv)_t = T(t,s)a + \int_s^t T_0(t,\tau)P_0(\tau)g(\tau, v_\tau)\, d\tau$$

$$- \int_t^{+\infty} \overline{T}(t,\tau)Q_0(\tau)g(\tau, v_\tau)\, d\tau$$

for $v \in \mathcal{X}^a$ and $t \geq s$. Clearly, Jv is continuous. Moreover,

$$(Jv)_s = a - \int_s^{+\infty} \overline{T}(s,\tau)Q_0(\tau)g(\tau, v_\tau)\, d\tau$$

and so $P(s)(Jv)_s = a$.

We want to show that $J(\mathcal{X}^a) \subset \mathcal{X}^a$ and that J is a contraction. For each $t \geq s$ and $v, w \in \mathcal{X}^a$, by Theorem 3.5 and (7.2) we obtain

$$\|(Jv)_t - (Jw)_t\| \leq N\delta \int_s^t e^{-\lambda(t-\tau)}\|v_\tau - w_\tau\|\, d\tau$$

$$+ N\delta \int_t^{+\infty} e^{\lambda(t-\tau)}\|v_\tau - w_\tau\|\, d\tau$$

$$\leq N\delta \int_s^t e^{-\lambda(t-\tau)}e^{-(\lambda - 2\delta N)(\tau-s)}\, d\tau\, |v - w|_{\mathcal{X}^a}$$

$$+ N\delta \int_t^{+\infty} e^{\lambda(t-\tau)}e^{-(\lambda - 2\delta N)(\tau-s)}\, d\tau\, |v - w|_{\mathcal{X}^a}$$

$$\leq N\delta e^{-\lambda(t-s)} \int_s^t e^{2\delta N(\tau-s)}\, d\tau |v-w|_{\mathcal{X}^a}$$

$$+ N\delta e^{\lambda(t-s)} \int_t^{+\infty} e^{-2(\lambda-\delta N)(\tau-s)}\, d\tau |v-w|_{\mathcal{X}^a}$$

$$\leq \frac{1}{2} e^{-(\lambda-2\delta N)(t-s)} |v-w|_{\mathcal{X}^a}$$

$$+ \frac{\delta N}{2(\lambda-\delta N)} e^{-(\lambda-2\delta N)(t-s)} |v-w|_{\mathcal{X}^a}$$

$$\leq \left(\frac{1}{2} + \frac{\delta N}{2(\lambda-\delta N)} \right) e^{-(\lambda-2\delta N)(t-s)} |v-w|_{\mathcal{X}^a}$$

and so

$$|Jv - Jw|_{\mathcal{X}^a} \leq \theta |v-w|_{\mathcal{X}^a},$$

where

$$\theta = \frac{1}{2} + \frac{\delta N}{2(\lambda-\delta N)}.$$

Taking δ sufficiently small so that $\theta < 2/3$ the operator J is a contraction. Moreover, again by Theorem 3.5, we have $|J0|_{\mathcal{X}^a} \leq N\|a\|$ and hence,

$$|Jv|_{\mathcal{X}^a} \leq |J0|_{\mathcal{X}^a} + |Jv - J0|_{\mathcal{X}^a}$$

$$\leq N\|a\| + \theta |v|_{\mathcal{X}^a}$$

$$\leq N\|a\| + \frac{2}{3} 3N\|a\| = 3N\|a\|.$$

This shows that $J(\mathcal{X}^a) \subset \mathcal{X}^a$ and so there exists a unique $v = v^{s,a} \in \mathcal{X}^a$ such that $Jv = v$.

We use the functions $v^{s,a}$ to construct the stable manifold. Taking $t = s$ in (7.12) we obtain

$$Q(s)v_s = -\int_s^{+\infty} \overline{T}(s,\tau) Q_0(\tau) g(\tau, v_\tau)\, d\tau.$$

Since we want that $V^s = \operatorname{graph} z$ for some $z \in \mathcal{S}^L$, we define a map

$$z \colon \left\{ (s,a) \in I \times C,\ a \in E(s) \right\} \to C$$

by

$$z(s,a) = -\int_s^{+\infty} \overline{T}(s,\tau) Q_0(\tau) g(\tau, v_\tau^{s,a})\, d\tau = Q(s) v_s^{s,a}. \tag{7.25}$$

Note that by construction,

$$v^{s,a} = v(\cdot, s, a + z(s,a)). \tag{7.26}$$

Now we show that $z \in \mathcal{S}^L$. Taking $\varphi \in C$ with $a = P(s)\varphi = 0$, the function $v = 0$ satisfies (7.12) (since $g(t,0) = 0$ for $t \in I$) and so $z(s,0) = 0$. Moreover, it follows from Lemma 7.1 taking

$$x(t) = \left\| v_t^{s,a} - v_t^{s,\overline{a}} \right\|, \quad \beta = N\|a - \overline{a}\|, \quad \gamma = \lambda \quad \text{and} \quad D = N$$

that

$$\left\| v_t^{s,a} - v_t^{s,\overline{a}} \right\| \leq 2N e^{-(\lambda - 2\delta N)(t-s)} \|a - \overline{a}\|.$$

This establishes inequality (7.24). On the other hand, by (7.25) we have

$$\|z(s,a) - z(s,\overline{a})\| \leq \delta N \int_s^{+\infty} e^{\lambda(s-\tau)} \left\| v_\tau^{s,a} - \overline{v}_\tau^{s,\overline{a}} \right\| d\tau$$

$$\leq 2\delta N^2 \int_s^{+\infty} e^{-2(\lambda - \delta N)(\tau - s)} \, d\tau \|a - \overline{a}\|$$

$$\leq \frac{\delta N^2}{\lambda - \delta N} \|a - \overline{a}\|.$$

Hence, taking δ sufficiently small we have $z \in \mathcal{S}^L$.

Finally, we show that graph $z = V^s$. First observe that given $(s,\varphi) \in$ graph z, we have $\varphi = a + z(s,a)$, where $a = P(s)\varphi$. Hence, by (7.25), $v^{s,a} = v(\cdot, s, \varphi)$ and since $v^{s,a}$ is bounded, we conclude that $(s,\varphi) \in V^s$. Conversely, if $(s,\varphi) \in V^s$, then by Propositions 7.2 and 7.3 there exists a function $v \in \mathcal{X}^a$ with $v_s = \varphi$, where $a = P(s)\varphi$. Moreover, in view of the uniqueness of solutions we have $v = v^{s,a}$. By (7.25) and (7.26) we obtain

$$z(s,a) = Q(s)v_s = Q(s)\varphi$$

and so

$$(s,\varphi) = (s, P(s)\varphi + Q(s)\varphi) = (s, a + z(s,a)) \in \text{graph } z.$$

This completes the proof of the theorem. $\qquad \square$

7.3 Lipschitz Parameter Dependence

In this section we consider perturbations depending on a parameter and we describe how the stable manifold constructed in Theorem 7.3 depends on this parameter. Namely, we consider perturbations of Eq. (2.12) of the form (7.5) for some continuous function $g \colon I \times C \times \Lambda \to \mathbb{R}^n$ satisfying (7.6).

As in Sec. 7.1, we assume that g is Lipschitz in the second variable uniformly on the parameter ρ and given $(s,\varphi,\rho) \in I \times C \times \Lambda$, we denote by $v(\cdot, s, \varphi, \rho)$ the unique solution of Eq. (7.5) with $v_s = \varphi$.

Theorem 7.4 (Parameter dependence). *Assume that Eq. (2.12) has an exponential dichotomy on an interval I containing $\mathbb{R}_0^+$ and that there exists $\delta > 0$ such that (7.7) and (7.8) hold for every $t \in I$, $u, v \in C$ and $\rho, \varsigma \in \Lambda$. Then, for any sufficiently small δ and each $\rho \in \Lambda$, the function $z = z^\rho \in S^L$ given by Theorem 7.3 satisfies*

$$\|w_t^{a,\rho} - w_t^{\overline{a},\rho}\| \leq 2N e^{-(\lambda - 2\delta N)(t-s)} \|a - \overline{a}\| \tag{7.27}$$

and

$$\|w_t^{a,\rho} - w_t^{a,\varsigma}\| \leq 4N e^{-(\lambda - 4\delta N)(t-s)} |\rho - \varsigma| \cdot \|a\| \tag{7.28}$$

for every $t, s \in I$ with $t \geq s$, $a, \overline{a} \in E(s)$ and $\rho, \varsigma \in \Lambda$, where

$$w_t^{a,\rho} = v_t(\cdot, s, a + z^\rho(s, a), \rho).$$

Proof. By Theorem 7.3, for each $\rho \in \Lambda$ there exists a function $z^\rho \in S^L$ such that the stable set of Eq. (7.5) is the graph of z^ρ. Moreover, property (7.27) holds.

It remains to establish property (7.28). Let

$$\alpha(t) = \|v_t - w_t\|,$$

where

$$v_t = v_t(\cdot, s, a + z^\rho(s, a), \rho)$$

and

$$w_t = w_t(\cdot, s, a + z^\varsigma(s, a), \varsigma).$$

We have

$$\begin{aligned}
|g(\tau, v_\tau, \rho) - g(\tau, w_\tau, \varsigma)| &\leq |g(\tau, v_\tau, \rho) - g(\tau, v_\tau, \varsigma)| \\
&\quad + |g(\tau, v_\tau, \varsigma) - g(\tau, w_\tau, \varsigma)| \\
&\leq 2\delta |\rho - \varsigma| \cdot \|v_\tau\| + \delta \alpha(\tau) \\
&\leq 4\delta N |\rho - \varsigma| \cdot \|a\| e^{-(\lambda - 2\delta N)(\tau - s)} + \delta \alpha(\tau),
\end{aligned}$$

using (7.14). Therefore, it follows from Proposition 7.2 that

$$\begin{aligned}
\alpha(t) &\leq 4\delta N^2 |\rho - \varsigma| \cdot \|a\| \int_s^t e^{-\lambda(t-\tau)} e^{-(\lambda - 2\delta N)(\tau - s)} \, d\tau \\
&\quad + \delta N \int_s^t e^{-\lambda(t-\tau)} \alpha(\tau) \, d\tau \\
&\quad + 4\delta N^2 |\rho - \varsigma| \cdot \|a\| \int_t^{+\infty} e^{\lambda(t-\tau)} e^{-(\lambda - 2\delta N)(\tau - s)} \, d\tau \\
&\quad + \delta N \int_t^{+\infty} e^{\lambda(t-\tau)} \alpha(\tau) \, d\tau
\end{aligned}$$

$$\leq \nu|\rho - \varsigma| \cdot \|a\| e^{-(\lambda - 2\delta N)(t-s)} + \delta N \int_s^t e^{-\lambda(t-\tau)} \alpha(\tau)\, d\tau$$

$$+ \delta N \int_t^{+\infty} e^{\lambda(t-\tau)} \alpha(\tau)\, d\tau$$

$$\leq \nu|\rho - \varsigma| \cdot \|a\| e^{-\mu(t-s)} + \delta N \int_s^t e^{-\mu(t-\tau)} \alpha(\tau)\, d\tau$$

$$+ \delta N \int_t^{+\infty} e^{\mu(t-\tau)} \alpha(\tau)\, d\tau$$

for $t \geq s$, where

$$\mu = \lambda - 2\delta N \quad \text{and} \quad \nu = 4\delta N^2 \left(\frac{1}{\delta N} + \frac{1}{\lambda - \delta N} \right).$$

By Theorem 7.3, the function α is bounded and continuous. Hence, it follows from Lemma 7.1 taking

$$x(t) = \alpha(t), \quad \beta = \nu|\rho - \varsigma| \cdot \|a\|, \quad \gamma = \mu \quad \text{and} \quad D = N$$

that

$$\alpha(t) \leq 2\nu|\rho - \varsigma| \cdot \|a\| e^{-(\mu - 2\delta N)} = 2\nu|\rho - \varsigma| \cdot \|a\| e^{-(\lambda - 4\delta N)}$$

for $t \geq s$, which establishes property (7.28). $\qquad\square$

7.4 Unstable Invariant Manifolds

In this section we consider the notion of unstable set of Eq. (7.1) and we obtain corresponding results to those in Sec. 7.2. In particular, we construct a Lipschitz unstable invariant manifold for Eq. (7.1) when Eq. (2.12) has an exponential dichotomy on an interval $I \subset \mathbb{R}$ containing $\mathbb{R}_0^-$ and g is a sufficiently small Lipschitz perturbation. We continue to assume that conditions 1 and 2 in Sec. 7.1 hold.

First we introduce the notion of unstable set.

Definition 7.2. The *unstable set* V^u of Eq. (7.1) is the set of all initial conditions $(s, \varphi) \in I \times C$ for which there exists a bounded solution v of the equation on the interval $(-\infty, s]$ with $v_s = \varphi$.

Proceeding in a similar manner to that in the proof of Proposition 7.1, one can establish the invariance of the unstable set.

Proposition 7.4. *If* $(s, \varphi) \in V^u$, *then* $(t, v_t) \in V^u$ *for all* $t \leq s$, *where* v *is the bounded solution in Definition 7.2.*

Moreover, the following result gives an alternative characterization of the unstable set when Eq. (2.12) has an exponential dichotomy.

Proposition 7.5. *Assume that Eq. (2.12) has an exponential dichotomy on an interval I containing $\mathbb{R}_0^-$. Then its unstable set is composed of the pairs $(s, \varphi) \in I \times C$ for which there exists a bounded continuous function $v \colon (-\infty, s] \to \mathbb{R}^n$ such that $v_s = \varphi$ and*

$$
\begin{aligned}
v_t = \overline{T}(t, s)Q(s)\varphi - \int_t^s \overline{T}(t, \tau)Q_0(\tau)g(\tau, v_\tau)\,d\tau \\
+ \int_{-\infty}^t T_0(t, \tau)P_0(\tau)g(\tau, v_\tau)\,d\tau
\end{aligned}
\tag{7.29}
$$

for every $t \le s$.

Proof. To the possible extent we follow the proof of Proposition 7.2. First assume that v is a bounded solution of Eq. (7.1) on the interval $(-\infty, s]$ with $v_s = \varphi$. By Theorem 3.6, the identities in (7.13) hold for $s \ge \tau \ge \sigma$. The second identity in (7.13) also holds for $\tau \le \sigma \le s$, provided that T is replaced (twice) by $\overline{T}$, that is,

$$
Q(\tau)v_\tau = \overline{T}(\tau, \sigma)Q(\sigma)v_\sigma - \int_\tau^\sigma \overline{T}(\tau, u)Q_0(u)g(u, v_u)\,du
\tag{7.30}
$$

for $\tau \le \sigma \le s$. Taking $\tau = t$, we write the first identity in (7.13) in the form

$$
P(t)v_t = T(t, \sigma)P(\sigma)v_\sigma + \int_\sigma^t T_0(t, u)P_0(u)g(u, v_u)\,du.
$$

By the first inequality in (3.23), since v is bounded we obtain

$$
\|\overline{T}(t, \sigma)P(\sigma)v_\sigma\| \le Ne^{-\lambda(t-\sigma)} \sup_{u \le s} \|v_u\| \to 0
$$

when $\sigma \to -\infty$. Hence,

$$
P(t)v_t = \int_{-\infty}^t T_0(t, u)P_0(u)g(u, v_u)\,du,
$$

which added to (7.30) with $\tau = t$ and $\sigma = s$ yields (7.29).

Now we assume that property (7.29) holds with $v_s = \varphi$. Taking $t = s$, we obtain

$$
v_s = Q(s)v_s + \int_{-\infty}^s T_0(s, u)P_0(u)g(u, v_u)\,du
$$

and hence,

$$
P(s)v_s = \int_{-\infty}^s T_0(s, u)P_0(u)g(u, v_u)\,du.
$$

Therefore, for each $t \leq s$ we have

$$T(s,t)v_t + \int_t^s T_0(s,u)X_0 g(u,v_u)\,du$$

$$= T(s,t)v_t + \int_t^s T_0(s,u)P_0(u)g(u,v_u)\,du + \int_t^s T(s,u)Q_0(u)g(u,v_u)\,du$$

$$= T(s,t)\overline{T}(t,s)Q(s)\varphi + T(s,t)\left(-\int_t^s \overline{T}(t,u)Q_0(u)g(u,v_u)\,du\right.$$

$$\left. + \int_{-\infty}^t T_0(t,u)P_0(u)g(u,v_u)\,du\right) + \int_t^s T_0(s,u)P_0(u)g(u,v_u)\,du$$

$$+ \int_t^s T(s,u)Q_0(u)g(u,v_u)\,du$$

$$= Q(s)\varphi + \int_{-\infty}^t T_0(s,u)P_0(u)g(u,v_u)\,du + \int_t^s T_0(s,u)P_0(u)g(u,v_u)\,du$$

$$= Q(s)\varphi + \int_{-\infty}^s T_0(s,u)P_0(u)g(u,v_u)\,du = v_s$$

and so v is a solution of Eq. (7.1).   □

Using the characterization of the unstable set in Proposition 7.5, one can show that all bounded solutions of Eq. (7.1) decay exponentially to the past.

Proposition 7.6 (Exponential decay). *Assume that Eq. (2.12) has an exponential dichotomy on an interval I containing $\mathbb{R}_0^-$ and that there exists $\delta > 0$ such that (7.2) holds for every $t \in I$ and $u,v \in C$. Then, for any sufficiently small δ, each bounded solution $v\colon (-\infty, s] \to \mathbb{R}^n$ of Eq. (7.1) with $v_s = \varphi$ satisfies*

$$\|v_t\| \leq 2N e^{-(\lambda - 2\delta N)(s-t)}\|Q(s)\varphi\| \tag{7.31}$$

for every $t \leq s$.

Proof. By Proposition 7.5, each bounded solution $v\colon (-\infty, s] \to \mathbb{R}^n$ of Eq. (7.1) with $v_s = \varphi$ satisfies (7.29). On the other hand, by Theorem 3.5 and (7.2) together with the fact that $g(t,0) = 0$, it follows from (7.29) that

$$\|v_t\| \leq N e^{-\lambda(t-s)}\|Q(s)\varphi\| + \delta N \int_t^s e^{-\lambda(\tau - t)}\|v_\tau\|\,d\tau$$

$$+ \delta N \int_{-\infty}^t e^{-\lambda(t-\tau)}\|v_\tau\|\,d\tau. \tag{7.32}$$

We also establish an auxiliary result.

Lemma 7.2. *Given $s \in I$, let $y \colon (-\infty, s] \to \mathbb{R}_0^+$ be a bounded continuous function such that*

$$x(t) \le \beta e^{-\gamma(s-t)} + \delta D \int_t^s e^{-\gamma(\tau-t)} x(\tau)\, d\tau + \delta D \int_{-\infty}^t e^{-\gamma(t-\tau)} x(\tau)\, d\tau$$

for every $t \le s$. Then, for any sufficiently small δ, we have

$$x(t) \le 2\beta e^{-(\gamma - 2\delta N)(s-t)} \quad \text{for } t \le s. \tag{7.33}$$

Proof of the lemma. Proceeding as in the proof of Lemma 7.1, one can show that $x(t) \le y(t)$, where y is the unique bounded continuous function satisfying the integral equation

$$y(t) = \beta e^{-\gamma(s-t)} + \delta D \int_t^s e^{-\gamma(\tau-t)} y(\tau)\, d\tau + \delta D \int_{-\infty}^t e^{-\gamma(t-\tau)} y(\tau)\, d\tau \tag{7.34}$$

for $t \le s$. We note that y is also a solution of the ordinary differential Eq. (7.19). Substituting $y(t) = y(s)e^{-\overline{\gamma}(s-t)}$ in (7.34) and taking $t = s$, we obtain

$$y(s) = \frac{\beta}{1 - \delta D/(\gamma + \overline{\gamma})},$$

provided that δ is sufficiently small. Therefore,

$$x(t) \le \frac{\beta}{1 - \delta D/(\gamma + \overline{\gamma})} e^{-\overline{\gamma}(s-t)}$$

for $t \le s$. Finally, proceeding as in Lemma 7.1, we find that (7.33) holds, again provided that δ is sufficiently small. $\qquad \square$

In view of (7.32), applying Lemma 7.2 taking

$$x(t) = \|v_t\|, \quad \beta = N\|Q(s)\varphi\|, \quad \gamma = \lambda \quad \text{and} \quad D = N,$$

we obtain inequality (7.31). $\qquad \square$

Now we formulate the Lipschitz unstable manifold theorem. Let $\mathcal{U}^L$ be the set of all continuous functions

$$w \colon \left\{ (s, b) \in I \times C : b \in F(s) \right\} \to C$$

such that for each $s \in I$:

(1) $w(s, 0) = 0$ and $w(s, F(s)) \subset E(s)$;
(2) for $b, \overline{b} \in F(s)$ we have

$$\|w(s, b) - w(s, \overline{b})\| \le \|b - \overline{b}\|.$$

For each function $w \in \mathcal{U}^L$ we consider its graph

$$\operatorname{graph} w = \left\{ (s, w(s,b) + b) : (s,b) \in I \times F(s) \right\} \subset I \times C.$$

Theorem 7.5 (Lipschitz unstable manifold). *Assume that Eq. (2.12) has an exponential dichotomy on an interval I containing $\mathbb{R}_0^-$ and that there exists $\delta > 0$ such that (7.2) holds for every $t \in I$ and $u, v \in C$. Then, for any sufficiently small δ, there exists a function $w \in \mathcal{U}^L$ such that $V^u = \operatorname{graph} w$. Moreover, for each initial condition $(s, w(s,b) + b) \in V^u$ there exists a unique bounded solution $v = v^{s,b}$ of Eq. (7.1) on $(-\infty, s]$ with $v_s = w(s,b) + b$, which satisfies*

$$\left\| v_t^{s,b} - v_t^{s,\overline{b}} \right\| \leq 2N e^{-(\lambda - 2\delta N)(s-t)} \|b - \overline{b}\|$$

for every $t, s \in I$ with $t \leq s$ and $b, \overline{b} \in F(s)$.

Proof. The argument is analogous to the one in the proof of Theorem 7.3 and so we only provide a brief sketch. Take $s \in I$ and $b \in F(s)$. We consider the set $\mathcal{Y}^b$ of all continuous functions $v \colon (-\infty, s] \to \mathbb{R}^n$ with $Q(s)v_s = b$ such that

$$\|v_t\| \leq 3N e^{-(\lambda - 2\delta N)(s-t)} \|b\|$$

for all $t \leq s$. One can easily verify that $\mathcal{Y}^b$ is a complete metric space when equipped with the norm

$$|v|_{\mathcal{Y}^b} := \sup\left\{ \|v_t\| e^{(\lambda - 2\delta N)(s-t)} : t \leq s \right\}.$$

We define an operator K on $\mathcal{Y}^b$ by

$$(Kv)_t = \overline{T}(t,s)b - \int_t^s \overline{T}(t,\tau)Q_0(\tau)g(\tau, v_\tau)\, d\tau$$

$$+ \int_{-\infty}^t T_0(t,\tau)P_0(\tau)g(\tau, v_\tau)\, d\tau$$

for $v \in \mathcal{Y}^b$ and $t \leq s$. In a similar manner to that in the proof of Theorem 7.3, provided that δ is sufficiently small, one can show that $K(\mathcal{Y}^b) \subset \mathcal{Y}^b$ and that K is a contraction. Hence, there exists a unique $v = v^{s,b} \in \mathcal{Y}^b$ such that $Kv = v$.

We use the functions $v^{s,b}$ to construct the unstable manifold. Taking $t = s$ in (7.29) we obtain

$$P(s)v_s = \int_{-\infty}^t T_0(t,\tau)P_0(\tau)g(\tau, v_\tau)\, d\tau.$$

Hence, in order that $V^u = \operatorname{graph} w$ for some $w \in \mathcal{U}^L$, we define a map

$$w \colon \big\{ (s,b) \in I \times C, \ b \in F(s) \big\} \to C$$

by

$$w(s,b) = \int_{-\infty}^{t} T_0(t,\tau) P_0(\tau) g(\tau, v_\tau^{s,b}) \, d\tau = Q(s) v_s^{s,b}.$$

Using Proposition 7.6, one can show in a similar manner to that in the proof of Theorem 7.3 that $w \in \mathcal{U}^L$ and $V^u = \operatorname{graph} w$. $\square$

7.5 Topological Conjugacies

In this section we establish a partial version of the Grobman–Hartman theorem in the theory of ordinary differential equations for Lipschitz perturbations of an exponential dichotomy on $\mathbb{R}$. We continue to assume that conditions 1 and 2 in Sec. 7.1 hold.

Assume that Eq. (2.12) has an exponential dichotomy on $\mathbb{R}$. We denote by $\mathcal{M}$ the set of all continuous functions

$$\eta \colon \big\{ (t,b) : t \in \mathbb{R}, b \in F(t) \big\} \to C$$

such that

$$\|\eta\|_\infty := \sup \big\{ \|\eta(t,b)\| : t \in \mathbb{R}, b \in F(t) \big\} < +\infty. \qquad (7.35)$$

Note that $\mathcal{M}$ is a Banach space with the norm $\|\cdot\|_\infty$ in (7.35). We also write

$$\eta^t = \eta(t,\cdot) \quad \text{and} \quad h^t = \operatorname{Id}_{F(t)} + \eta^t.$$

We continue to denote by $T(t,s)$ the evolution family associated with the equation $v' = L(t)v_t$ and we write the solutions of Eq. (7.1) in the form

$$v_t = R(t,s)(v_s) \quad \text{and } t \geq s.$$

Theorem 7.6. *Assume that Eq. (2.12) has an exponential dichotomy on $\mathbb{R}$ and that there exists $\delta > 0$ such that*

$$|g(t,u) - g(t,v)| \leq \delta \min\{1, \|u - v\|\} \qquad (7.36)$$

for every $t \in \mathbb{R}$ and $u, v \in C$. If δ is sufficiently small, then there exists a unique $\eta \in \mathcal{M}$ such that

$$h^t \circ T(t,s) = R(t,s) \circ h^s \quad \text{on } F(s) \qquad (7.37)$$

for every $t, s \in \mathbb{R}$ with $t \geq s$. Moreover, each map h^t is one-to-one.

Proof. We define an operator $F \colon \mathcal{M} \to \mathcal{M}$ by

$$F(\eta)(t,b) = \int_{-\infty}^{t} T_0(t,\tau) P_0(\tau) f_\eta(\tau, \overline{T}(\tau,t)b) \, d\tau$$
$$- \int_{t}^{+\infty} \overline{T}(t,\tau) Q_0(\tau) f_\eta(\tau, T(\tau,t)b) \, d\tau \qquad (7.38)$$

for $\eta \in \mathcal{M}$, $t \in \mathbb{R}$ and $b \in F(t)$, where

$$f_\eta(t,b) = g(t, b + \eta(t,b)).$$

By (3.24) and (7.36) together with the fact that $g(t,0) = 0$, we have

$$\int_{-\infty}^{t} \|T_0(t,\tau) P_0(\tau)\| \cdot |f_\eta(\tau, \overline{T}(\tau,t)b)| \, d\tau \le N\delta \int_{-\infty}^{t} e^{-\lambda(t-\tau)} \, d\tau = \frac{N\delta}{\lambda}$$

and

$$\int_{t}^{+\infty} \|\overline{T}(t,\tau) Q_0(\tau)\| \cdot |f_\eta(\tau, T(\tau,t)b)| \, d\tau \le N\delta \int_{t}^{+\infty} e^{-\lambda(\tau-t)} \, d\tau = \frac{N\delta}{\lambda}.$$

This shows that $F(\eta)$ is well defined.

Lemma 7.3. *For any sufficiently small δ, there exists a unique $\eta \in \mathcal{M}$ such that $F(\eta) = \eta$.*

Proof of the lemma. It suffices to show that the operator F is a contraction. For each $\eta, \xi \in \mathcal{M}$ we have

$$F(\eta)(t,b) - F(\xi)(t,b)$$
$$= \int_{-\infty}^{t} T_0(t,\tau) P_0(\tau) \big(f_\eta(\tau, \overline{T}(\tau,t)b) - f_\xi(\tau, \overline{T}(\tau,t)b) \big) \, d\tau$$
$$- \int_{t}^{+\infty} \overline{T}(t,\tau) Q_0(\tau) \big(f_\eta(\tau, T(\tau,t)b) - f_\xi(\tau, T(\tau,t)b) \big) \, d\tau.$$

It follows from (7.36) that

$$|f_\eta(\tau, \overline{T}(\tau,t)b) - f_\xi(\tau, \overline{T}(\tau,t)b)| \le \delta \|\eta(\tau, \overline{T}(\tau,t)b) - \xi(\tau, \overline{T}(\tau,t)b)\|$$

for $\tau \le t$ and

$$|f_\eta(\tau, T(\tau,t)b) - f_\xi(\tau, T(\tau,t)b)| \le \delta \|\eta(\tau, T(\tau,t)b) - \xi(\tau, T(\tau,t)b)\|$$

for $\tau \ge t$. Hence, using (3.24) we obtain

$$\|F(\eta)(t,b) - F(\xi)(t,b)\|$$
$$\le N\delta \int_{-\infty}^{t} e^{-\lambda(t-\tau)} \|\eta(\tau, \overline{T}(\tau,t)b) - \xi(\tau, \overline{T}(\tau,t)b)\| \, d\tau$$
$$+ N\delta \int_{t}^{+\infty} e^{-\lambda(\tau-t)} \|\eta(\tau, T(\tau,t)b) - \xi(\tau, T(\tau,t)b)\| \, d\tau$$

$$\leq N\delta \int_{-\infty}^{t} e^{-\lambda(t-\tau)} \|\eta - \xi\|_\infty \, d\tau + N\delta \int_{t}^{+\infty} e^{-\lambda(\tau-t)} \|\eta - \xi\|_\infty \, d\tau$$

$$= \frac{2N\delta}{\lambda} \|\eta - \xi\|_\infty$$

and so

$$\|F(\eta) - F(\xi)\|_\infty \leq \frac{2N\delta}{\lambda} \|\eta - \xi\|_\infty.$$

Therefore, for δ sufficiently small the operator F is a contraction.   □

Now we show that the unique fixed point of F given by Lemma 7.3 is also the unique solution of problem (7.37).

Lemma 7.4. $F(\eta) = \eta$ *if and only if property* (7.37) *holds.*

Proof of the lemma. First assume that $F(\eta) = \eta$ and take $t, s \in \mathbb{R}$ with $t \geq s$. By (7.38), for each $b \in F(t)$ we have

$$Q(t)\eta^t(b) = -\int_{t}^{+\infty} \overline{T}(t,\tau) Q_0(\tau) g(\tau, h^\tau(T(\tau,t)b)) \, d\tau \qquad (7.39)$$

for $b \in F(t)$. Since

$$Q(s)\eta^s(c) = -\int_{s}^{+\infty} \overline{T}(s,\tau) Q_0(\tau) g(\tau, h^\tau(T(\tau,s)c)) \, d\tau \qquad (7.40)$$

for $c = \overline{T}(s,t)b \in F(s)$ and

$$\overline{T}(s,t)Q(t)\eta^t(b) = -\int_{t}^{+\infty} \overline{T}(s,\tau) Q_0(\tau) g(\tau, h^\tau(T(\tau,t)b)) \, d\tau,$$

we obtain

$$\overline{T}(s,t)Q(t)\eta^t(b) = Q(s)\eta^s(\overline{T}(s,t)b)$$
$$+ \int_{s}^{t} \overline{T}(s,\tau) Q_0(\tau) g(\tau, h^\tau(\overline{T}(\tau,t)b)) \, d\tau$$

or, equivalently,

$$Q(t)\eta^t(b) = T(t,s)Q(s)\eta^s(\overline{T}(s,t)b)$$
$$+ \int_{s}^{t} T(t,\tau) Q_0(\tau) g(\tau, h^\tau(\overline{T}(\tau,t)b)) \, d\tau. \qquad (7.41)$$

On the other hand,

$$P(t)\eta^t(b) = \int_{-\infty}^{t} T_0(t,\tau) P_0(\tau) g(\tau, h^\tau(\overline{T}(\tau,t)b)) \, d\tau. \qquad (7.42)$$

Since

$$P(s)\eta^s(c) = \int_{-\infty}^s T_0(s,\tau)P_0(\tau)g(\tau, h^\tau(\overline{T}(\tau,s)c))\, d\tau$$

for $c = \overline{T}(s,t)b \in F(s)$ and

$$T(t,s)P(s)\eta^s(\overline{T}(s,t)b) = \int_{-\infty}^s T_0(t,\tau)P_0(\tau)g(\tau, h^\tau(\overline{T}(\tau,t)b))\, d\tau,$$

we obtain

$$P(t)\eta^t(b) = T(t,s)P(s)\eta^s(\overline{T}(s,t)b)$$
$$+ \int_s^t T_0(t,\tau)P_0(\tau)g(\tau, h^\tau(\overline{T}(\tau,t)b))\, d\tau. \tag{7.43}$$

Adding (7.41) and (7.43) yields the identity

$$\eta^t(b) = T(t,s)\eta^s(\overline{T}(s,t)b) + \int_s^t T(t,\tau)X_0 g(\tau, h^\tau(\overline{T}(\tau,t)b))\, d\tau.$$

Finally, since $h^t = \mathrm{Id}_{F(t)} + \eta^t$ we obtain

$$h^t(b) = T(t,s)h^s(\overline{T}(s,t)b) + \int_s^t T(t,\tau)X_0 g(\tau, h^\tau(\overline{T}(\tau,t)b))\, d\tau. \tag{7.44}$$

On the other hand, by the variation of constants formula and the definition of $R(t,s)$, we have

$$R(t,s)(v_s) = T(t,s)v_s + \int_s^t T(t,\tau)X_0 g(\tau, R(\tau,s)(v_s))\, d\tau$$

for $t \geq s$. Comparing with (7.44) we conclude that

$$R(t,s)(h^s(c)) = h^t(T(t,s)c)$$

for $t \geq s$ and $c \in F(s)$, which establishes (7.37).

Now assume that (7.37) holds. Then

$$h^t(T(t,s)c) = T(t,s)h^s(c) + \int_s^t T(t,\tau)X_0 g(\tau, h^\tau(T(\tau,s)c))\, d\tau,$$

which is equivalent to (7.44). Projecting onto the stable and unstable directions, it follows from Theorem 3.6 that identities (7.41) and (7.43) hold. The first is equivalent to

$$\overline{T}(s,t)Q(t)\eta^t(T(t,s)c) = Q(s)\eta^s(c) + \int_s^t \overline{T}(s,\tau)Q_0(\tau)g(\tau, h^\tau(T(\tau,s)c))\, d\tau.$$

Letting $t \to +\infty$ yields (7.40) and since s is arbitrary, we also obtain (7.39). On the other hand, letting $s \to -\infty$ in (7.43) yields (7.42). Finally, adding (7.39) and (7.42) we find that $F(\eta) = \eta$. This completes the proof of the lemma. $\qquad\square$

Lemma 7.5. *The map h^t is one-to-one for each $t \in \mathbb{R}$.*

Proof of the lemma. Take $c, \bar{c} \in F(s)$ such that $h^s(c) = h^s(\bar{c})$. It follows from (7.37) that

$$h^t(T(t,s)c) = h^t(T(t,s)\bar{c})$$

or, equivalently,

$$T(t,s)(c - \bar{c}) = -\left[\eta^t(T(t,s)c) - \eta^t(T(t,s)\bar{c}) \right], \qquad (7.45)$$

for $t \geq s$. By (3.23) we have

$$\|T(t,s)(c - \bar{c})\| \geq \frac{\|c - \bar{c}\|}{\|\overline{T}(s,t)\|} \geq \frac{1}{N} e^{\lambda(t-s)} \|c - \bar{c}\|.$$

Hence, since $\lambda > 0$, if $c \neq \bar{c}$, then the function $t \mapsto T(t,s)(c - \bar{c})$ is unbounded for $t \geq s$. On the other hand, the right-hand side of (7.45) is bounded by $2\|\eta\|_\infty$, which shows that $c = \bar{c}$. $\square$

The theorem follows now readily from the former lemmas. $\square$

Chapter 8

Smooth Invariant Manifolds

This chapter is dedicated to the construction of smooth stable and unstable invariant manifolds for any sufficiently small smooth perturbation of an exponential dichotomy. In view of the uniqueness of the Lipschitz invariant manifolds constructed in Chap. 7 it remains to show that the function of which the invariant manifold is a graph has the required regularity properties. More precisely, we establish the existence of stable and unstable invariant manifolds for C^1 perturbations and C^1 perturbations with Lipschitz derivative. In each case we show that the manifolds have the same regularity as the perturbation, thus obtaining optimal results. We also discuss briefly the existence of C^k invariant manifolds when the perturbation is of class C^k. We omit the proofs of these higher smoothness results since they require material that is only described at length in Chap. 9 in connection with the construction of center manifolds.

8.1 Smooth Stable Invariant Manifolds

In this section we consider again nonlinear perturbations of Eq. (2.12) and we establish the existence of a smooth stable invariant manifold for Eq. (7.1) when the linear equation $v' = L(t)v_t$ has an exponential dichotomy on an interval $I \subset \mathbb{R}$ containing $\mathbb{R}_0^+$ and g is a sufficiently small C^1 perturbation. We assume that:

(1) $L(t) \colon C \to \mathbb{R}^n$, for $t \in I$, are bounded linear operators such that the map $(t, v) \mapsto L(t)v$ is of class C^1 and condition (2.22) holds;
(2) $g \colon I \times C \to \mathbb{R}^n$ is of class C^1, and $g(t, 0) = 0$ and $\partial g(t, 0) = 0$ for all $t \in I$, where ∂ denotes the partial derivative with respect to the second variable.

Given $(s, \varphi) \in I \times C$, we continue to denote by $v(\cdot, s, \varphi)$ the unique solution of the Eq. (7.1) with $v_s = \varphi$.

The stable manifold is obtained as a graph of a function of class C^1 in the second variable. In Theorem 7.3 we have already shown that the stable set is the graph of a Lipschitz function z in the second variable (and since V^s is a graph, z is unique: two functions coincide if and only if their graphs coincide). Therefore, what remains to show in the present setting is that this function is of class C^1 in the second variable. Let $\mathcal{S}^1$ be the set of all functions $z \in \mathcal{S}^L$ (see Sec. 7.2) of class C^1 in the second variable such that $\partial z(s, 0) = 0$ for $s \in I$ (where ∂ denotes the partial derivative with respect to the second variable). The graph of a function $z \in \mathcal{S}^1$ is given by (7.23).

Theorem 8.1 (Smooth stable manifold). *Assume that Eq. (2.12) has an exponential dichotomy on an interval I containing $\mathbb{R}_0^+$ and that there exists $\delta > 0$ such that*

$$\|\partial g(t, v)\| \le \delta \tag{8.1}$$

for every $t \in I$ and $v \in C$. Then, for any sufficiently small δ, the function $z \in \mathcal{S}^L$ given by Theorem 7.3 is in $\mathcal{S}^1$.

Proof. To show that $z \in \mathcal{S}^L$ we will use the following result of Henry [59]. Our proof is based on [54].

Lemma 8.1. *Given Banach spaces Y and Z, let $f \colon A \to Z$ be a Lipschitz function on some open ball $A \subset Y$. Then f is of class C^1 if and only if for each $x \in A$ we have*

$$|f(y + h) - f(y) - f(x + h) + f(x)| = o(|h|) \tag{8.2}$$

when $(y, h) \to (x, 0)$.

Proof of the lemma. Clearly, if f is a C^1 function, then (8.2) holds.

Now assume that (8.2) holds. First observe that if f is differentiable on A, then df is continuous. Indeed, given $x \in A$ and $\varepsilon > 0$, there exists $\delta > 0$ such that if $|y - x| < \delta$ and $|h| < \delta$, then by (8.2) we have

$$\left| \frac{f(y + hz) - f(y)}{h} - \frac{f(x + hz) - f(x)}{h} \right| \le \varepsilon \tag{8.3}$$

for every $z \in Z$ with $|z| = 1$. Letting $h \to 0$ we conclude that

$$|d_y f z - d_x f z| \le \varepsilon$$

uniformly in z with $|z| = 1$, which shows that the derivative is continuous. Hence, it suffices to show that f is differentiable on A.

First assume that $Y = Z = \mathbb{R}$. Since f is Lipschitz, it is differentiable almost everywhere. Take $x \in A$ at which f is differentiable. It follows from (8.3) with $|z| = 1$ and $|y - x| < \delta$ that

$$f'(x) - \varepsilon \leq \limsup_{h \to 0} \frac{f(y + h) - f(y)}{h} \leq f'(x) + \varepsilon$$

and

$$f'(x) - \varepsilon \leq \liminf_{h \to 0} \frac{f(y + h) - f(y)}{h} \leq f'(x) + \varepsilon.$$

Therefore,

$$-2\varepsilon \leq \limsup_{h \to 0} \frac{f(y + h) - f(y)}{h} - \liminf_{h \to 0} \frac{f(y + h) - f(y)}{h} \leq 2\varepsilon.$$

Letting $\varepsilon \to 0$ we conclude that f is differentiable at y.

Now assume that $Y = \mathbb{R}$ and that Z is an arbitrary Banach space. Take $F \in Z^*$, where Z^* is the dual of Z. Then $F \circ f \colon A \to \mathbb{R}$ is a Lipschitz function with Lipschitz constant

$$\mathrm{Lip}(F \circ f) \leq \|F\| \, \mathrm{Lip}\, f.$$

In view of the former discussion when $Z = \mathbb{R}$, we know that $F \circ f$ is a C^1 function, which thus satisfies (8.2). Moreover, for each $x \in \mathbb{R}$ we have

$$|(F \circ f)'(x)| \leq \|F\| \, \mathrm{Lip}\, f. \tag{8.4}$$

Now we define a map $D(x) \in Z^{**}$ by

$$D(x)F = (F \circ f)'(x).$$

Clearly, $D(x)$ is linear and in view of (8.4) it is also continuous. For each $y \in A \subset \mathbb{R}$ and $F \in Z$ with $\|F\| \leq 1$, we have

$$\int_y^{y+h} D(x)F \, dx = \int_y^{y+h} (F \circ f)'(x) \, dx = Ff(y + h) - Ff(y)$$

and so

$$Ff(y + h) - Ff(y) - hD(y)F = \int_y^{y+h} \big(D(x) - D(y)\big)F \, dx.$$

Hence, since $D(x)$ is continuous, we obtain

$$F\left(\frac{f(y + h) - f(y)}{h}\right) - D(y)F = \frac{1}{h} \int_y^{y+h} \big(D(x) - D(y)\big)F \, dx \to 0$$

when $h \to 0$, uniformly in F with $\|F\| \leq 1$. Now let $\iota \colon Z \to Z^{**}$ be the canonical inclusion. By definition, we have

$$\iota\left(\frac{f(y + h) - f(y)}{h}\right)F = F\left(\frac{f(y + h) - f(y)}{h}\right)$$

and so

$$\left\| \iota\left(\frac{f(y+h) - f(y)}{h}\right) - D(y) \right\| \to 0 \tag{8.5}$$

when $h \to 0$. Since ι is continuous and one-to-one, it follows from (8.5) that the limit of

$$\frac{f(y+h) - f(y)}{h} \in Z$$

when $h \to 0$ exists (and takes values in Z). Hence, f is differentiable at y.

Finally, let Y and Z be arbitrary Banach spaces. By the former discussion when $Y = \mathbb{R}$ and Z is an arbitrary Banach space, we know that given $y, h \in Y$ with $y \in A$, the map $t \mapsto f(y + th)$ is of class C^1 in a sufficiently small neighborhood of the origin. In particular, the Gâteaux derivative

$$df(y, h) = \lim_{t \to 0} \frac{f(y + th) - f(y)}{t}$$

exists. Since $t \mapsto f(y+th)$ is of class C^1, it satisfies (8.2) and so $df(y, h) \to df(x, h)$ in Z when $y \to x \in A$, uniformly in h with $|h| \leq 1$. Indeed, given $x \in A$ and $\varepsilon \geq 0$, there exists $\delta > 0$ such that if $|y - x| < \delta$ and $|th| < \delta$, then

$$|f(y + th) - f(y) - f(x + th) + f(x)| \leq \varepsilon |th|.$$

Hence, for $|h| \leq 1$ we obtain

$$\left| \frac{f(y+th) - f(y)}{t} - \frac{f(x+th) - f(x)}{t} \right| \leq \varepsilon$$

and letting $t \to 0$ yields the inequality

$$|df(y, h) - df(x, h)| \leq \varepsilon.$$

This establishes the continuity of the Gâteaux derivative $df(y, h)$ in y, uniformly in h with $|h| \leq 1$. Moreover, since f is Lipschitz, the map $h \mapsto df(y, h)$ is also continuous. Clearly,

$$df(y, ch) = c\, df(y, h)$$

for $c \in \mathbb{R}$. Hence, to complete the proof of the lemma, it remains to show that the Gâteaux derivative is additive in h. Given $h_1, h_2 \in Y$, we have

$$\begin{aligned}
df(y, h_1 + h_2) &= \lim_{t \to 0} \frac{f(y + t(h_1 + h_2)) - f(y)}{t} \\
&= \lim_{t \to 0} \frac{f(y + t(h_1 + h_2)) - f(y + th_1) - f(y + th_2) + f(y)}{t} \\
&\quad + \lim_{t \to 0} \frac{f(y + th_1) - f(y)}{t} + \lim_{t \to 0} \frac{f(y + th_2) - f(y)}{t} \\
&= \lim_{t \to 0} \frac{f(y + t(h_1 + h_2)) - f(y + th_1) - f(y + th_2) + f(y)}{t} \\
&\quad + df(y, h_1) + df(y, h_2).
\end{aligned}$$

In view of (8.2) the last limit vanishes. Indeed, it follows from (8.2) that given $y \in A$ and $\varepsilon > 0$, there exists $\delta > 0$ such that if $|t| < \delta$, then

$$|f(y + th_1 + th_2) - f(y + th_1) - f(y + th_2) + f(y)| \le \varepsilon|th_2|.$$

Hence,

$$\left| \frac{f(y + th_1 + th_2) - f(y + th_1) - f(y + th_2) + f(y)}{t} \right| \le \varepsilon|h_2|$$

for all sufficiently small t and since ε is arbitrary, we conclude that

$$\lim_{t \to 0} \frac{f(y + t(h_1 + h_2)) - f(y + th_1) - f(y + th_2) + f(y)}{t} = 0.$$

Therefore, $h \mapsto df(y, h)$ is the Fréchet derivative of f at y and the proof of the lemma is complete. $\qquad\square$

We proceed with the proof of the theorem. Let $z \in S^L$ be the function given by Theorem 7.3. Notice that the map $E(s) \ni a \mapsto z(s, a)$ is of class C^1 if and only if

$$a \mapsto a + z(s, a) = v_s(\cdot, s, a + z(s, a))$$

is of class C^1. Hence, in view of Lemma 8.1 and writing

$$w_t^a = v_t(\cdot, s, a + z(s, a)),$$

it suffices to show that

$$\|w_t^{b+h} - w_t^b - w_t^{a+h} + w_t^a\| = o(\|h\|)$$

when $(b, h) \to (a, 0)$, for each $s \in I$, $t \ge s$ and $a \in E(s)$ (in fact it suffices to take $t = s$ but for the argument we need to consider an arbitrary $t \ge s$).

We define

$$\Upsilon_t(a, b, h) = w_t^{b+h} - w_t^b - w_t^{a+h} + w_t^a.$$

By Taylor's formula we have

$$g(\tau, w_\tau^{a+h}) = g(\tau, w_\tau^a + (w_\tau^{a+h} - w_\tau^a))$$
$$= g(\tau, w_\tau^a) + \partial g(\tau, w_\tau^a)(w_\tau^{a+h} - w_\tau^a) + \Delta(\tau, a, h),$$

where

$$\Delta(\tau, a, h) = \int_0^1 \left(\partial g(\tau, w_\tau^a + t(w_\tau^{a+h} - w_\tau^a)) - \partial g(\tau, w_\tau^a) \right)(w_\tau^{a+h} - w_\tau^a) \, dt.$$

Therefore, one can easily verify that

$$G(\tau) := g(\tau, w_\tau^{b+h}) - g(\tau, w_\tau^b) - g(\tau, w_\tau^{a+h}) + g(\tau, w_\tau^a)$$
$$= \partial g(\tau, w_\tau^a)\Upsilon_\tau(a, b, h) + \left(\partial g(\tau, w_\tau^b) - \partial g(\tau, w_\tau^a) \right)(w_\tau^{b+h} - w_\tau^b)$$
$$+ \Delta(\tau, b, h) - \Delta(\tau, a, h).$$

Since

$$w_t^a = T(t,s)a + \int_s^t T_0(t,\tau)P_0(\tau)g(\tau, w_\tau^a)\, d\tau$$

$$- \int_t^{+\infty} \overline{T}(s,\tau)Q_0(\tau)g(\tau, w_\tau^a)\, d\tau$$

for $t \geq s$, we obtain

$$\Upsilon_t(a,b,h) = \int_s^t T_0(t,\tau)P_0(\tau)G(\tau)\, d\tau - \int_t^{+\infty} \overline{T}(s,\tau)Q_0(\tau)G(\tau)\, d\tau$$

$$= \int_s^t T_0(t,\tau)P_0(\tau)\partial g(\tau, w_\tau^a)\Upsilon_\tau(a,b,h)\, d\tau$$

$$+ \int_s^t T_0(t,\tau)P_0(\tau)\big(\partial g(\tau, w_\tau^b) - \partial g(\tau, w_\tau^a)\big)(w_\tau^{b+h} - w_\tau^b)\, d\tau$$

$$+ \int_s^t T_0(t,\tau)P_0(\tau)\big(\Delta(\tau,b,h) - \Delta(\tau,a,h)\big)\, d\tau$$

$$- \int_t^{+\infty} \overline{T}(s,\tau)Q_0(\tau)\partial g(\tau, w_\tau^a)\Upsilon_\tau(a,b,h)\, d\tau$$

$$- \int_t^{+\infty} \overline{T}(s,\tau)Q_0(\tau)\big(\partial g(\tau, w_\tau^b) - \partial g(\tau, w_\tau^a)\big)(w_\tau^{b+h} - w_\tau^b)\, d\tau$$

$$- \int_t^{+\infty} \overline{T}(s,\tau)Q_0(\tau)\big(\Delta(\tau,b,h) - \Delta(\tau,a,h)\big)\, d\tau.$$

$$(8.6)$$

Now we estimate each integral I_i, for $i = 1, 2, 3, 4, 5, 6$, on the right-hand side of (8.6). Using inequalities (3.24) and (8.1), we obtain

$$|I_1| \leq \delta N \int_s^t e^{-\lambda(t-\tau)} \sup_{\tau \geq s}\|\Upsilon_\tau(a,b,h)\|\, d\tau \leq \frac{\delta N}{\lambda} \sup_{t \geq s}\|\Upsilon_t(a,b,h)\|$$

and

$$|I_4| \leq \frac{\delta N}{\lambda} \sup_{t \geq s}\|\Upsilon_t(a,b,h)\|.$$

Moreover, by Theorem 7.3 we have

$$\|w_t^{b+h} - w_t^b\| \leq 2N\|h\|e^{-(\lambda - 2\delta N)(t-s)} \tag{8.7}$$

for any $t \geq s$ and $b \in E(s)$. In particular, $\sup_{t \geq s}\|\Upsilon_t(a,b,h)\|$ is finite. Letting

$$S_1(a,b) = \sup_{t \geq s} \left(\big\|\partial g(t, w_t^b) - \partial g(t, w_t^a)\big\|e^{-(\lambda - 2\delta N)(t-s)}\right),$$

we obtain

$$|I_2| \leq 2N^2 \|h\| S_1(a,b) \int_s^t e^{-\lambda(t-\tau)} \, d\tau \leq \frac{2N^2}{\lambda} \|h\| S_1(a,b).$$

Analogously,

$$|I_5| \leq \frac{2N^2}{\lambda} \|h\| S_1(a,b).$$

Finally, let

$$S_2(a,b,h) = \sup_{\tau \geq s} \left(\int_0^1 \left(\tilde{g}(\tau,a) + \tilde{g}(\tau,b) \right) dt \, e^{-(\lambda-2\delta N)(\tau-s)} \right),$$

where

$$\tilde{g}(\tau,c) = \left\| \partial g(\tau, w_\tau^c + t(w_\tau^{c+h} - w_\tau^c)) - \partial g(\tau, w_\tau^c) \right\|.$$

Proceeding as above, it follows from the definitions of $\Delta(\tau, b, h)$ and $\Delta(\tau, a, h)$ together with (8.7) that

$$|I_3| \leq 2N^2 \|h\| S_2(a,b,h) \int_s^t e^{-\lambda(t-\tau)} \, d\tau \leq \frac{2N^2}{\lambda} \|h\| S_2(a,b,h)$$

and, similarly,

$$|I_6| \leq \frac{2N^2}{\lambda} \|h\| S_2(a,b,h).$$

Therefore, by (8.6), taking δ sufficiently small so that $\delta N \leq \lambda/4$ we obtain the upper bound

$$\sup_{t \geq s} \|\Upsilon_t(a,b,h)\| \leq \frac{8N^2}{\lambda} \|h\| (S_1(a,b) + S_2(a,b,h)). \tag{8.8}$$

In order to estimate the right-hand side of (8.8), we separate the suprema in $S_1(a,b)$ and $S_2(a,b,h)$ over $[s, +\infty)$ into suprema over $[s,T]$ and $[T, +\infty)$ for some sufficiently large T. Namely, given $\varepsilon > 0$, take $T > s$ such that

$$\delta e^{-(\lambda-2\delta N)(T-s)} < \varepsilon.$$

Then

$$\sup_{t \geq T} \left(\left\| \partial g(t, w_t^b) - \partial g(t, w_t^a) \right\| e^{-(\lambda-2\delta N)(t-s)} \right)$$
$$+ \sup_{\tau \geq T} \left(\int_0^1 \left(\tilde{g}(\tau,a) + \tilde{g}(\tau,b) \right) dt \, e^{-(\lambda-2\delta N)(\tau-s)} \right) \leq 6\varepsilon. \tag{8.9}$$

Now we consider the interval $[s, T]$. Since the composition $(t, a) \mapsto \partial g(t, w_t^a)$ is continuous (see Theorem 7.3), given $t \in [s, T]$, there exists $\delta_t > 0$ such that

$$\left\| \partial g(p, w_p^c) - \partial g(t, w_t^a) \right\| < \varepsilon$$

for any $p \in [s, T]$ and $c \in E(s)$ with $|p - t| < \delta_t$ and $\|c - a\| < \delta_t$. Note that

$$\left\| \partial g(p, w_p^c) - \partial g(q, w_q^d) \right\| < 2\varepsilon \tag{8.10}$$

for any other $q \in [s, T]$ and $d \in E(s)$ with $|q - t| < \delta_t$ and $\|d - a\| < \delta_t$. The open intervals $I_t = (t - \delta_t, t + \delta_t)$ cover the compact set $[s, T]$ and so one can consider a finite subcover $I_{t_1}, \ldots, I_{t_n}$. Letting $\delta = \min\{\delta_{t_i} : i = 1, \ldots, n\}$, it follows from (8.10) with $p = q = t$ that

$$\left\| \partial g(t, w_t^c) - \partial g(t, w_t^d) \right\| < 2\varepsilon \tag{8.11}$$

for any $t \in [s, T]$ and $c, d \in E(s)$ such that $\|c - a\| < \delta$ and $\|d - a\| < \delta$. Now take $b, h \in E(s)$ with $\|b - a\| < \delta/2$ and $\|h\| < \delta/2$. It follows readily from (8.11) that

$$\sup_{t \in [s,T]} \left\| \partial g(t, w_t^b) - \partial g(t, w_t^a) \right\| + \sup_{\tau \in [s,T]} \left(\int_0^1 \left(\tilde{g}(\tau, a) + \tilde{g}(\tau, b) \right) dt \right) \le 6\varepsilon.$$

Together with (8.8) and (8.9), this implies that

$$\sup_{t \ge s} \| \Upsilon_t(a, b, h) \| \le \frac{96 N^2}{\lambda} \|h\| \varepsilon.$$

Since ε is arbitrary, we obtain

$$\lim_{(b,h) \to (a,0)} \frac{1}{\|h\|} \sup_{t \ge s} \| \Upsilon_t(a, b, h) \| = 0$$

and applying Lemma 8.1, we find that $E(s) \ni a \mapsto z(s, a)$ is a C^1 function. Moreover, by (7.25) and the Leibniz rule, we have

$$\partial z(s, 0) = -\int_s^{+\infty} \overline{T}(s, \tau) Q_0(\tau) \partial g(\tau, v_\tau^{s,0}) \frac{\partial v_\tau^{s,a}}{\partial a} \Big|_{a=0} d\tau = 0$$

since $v_\tau^{s,0} = 0$ and $\partial g(\tau, 0) = 0$ for all τ (by hypothesis). Hence, $z \in \mathcal{S}^1$. This completes the proof of the theorem. $\qquad \square$

8.2   Graphs with Lipschitz Derivatives

In this section we consider again nonlinear perturbations of Eq. (2.12) and we establish the existence of a smooth stable invariant manifold for Eq. (7.1) when $v' = L(t)v_t$ has an exponential dichotomy on $\mathbb{R}_0^+$ and g is a sufficiently small C^1 perturbation with Lipschitz derivative. In comparison to Theorem 8.1, we show, in addition, that for the unique function $z \in \mathcal{S}^L$ given by Theorem 7.3 each function $s \mapsto z(s,a)$ has a Lipschitz derivative. Theorem 8.1 only shows that this function is of class C^1, although without requiring that the derivative of g is Lipschitz. We emphasize that both Theorem 8.1 and the result in the present section establish the optimal regularity of the function $s \mapsto z(s,a)$ in the corresponding setting. We continue to assume that conditions 1 and 2 in Sec. 8.1 hold.

We assume that Eq. (2.12) has an exponential dichotomy on an interval $I \subset \mathbb{R}$ containing $\mathbb{R}_0^+$. Let $\mathcal{S}^{1+L}$ be the set of all functions $z \in \mathcal{S}^1$ (see Sec. 8.1) such that

$$\|\partial z(s,a) - \partial z(s,\bar{a})\| \leq \|a - \bar{a}\|$$

for every $s \in I$ and $a, \bar{a} \in E(s)$ (where ∂ continues to denote the partial derivative with respect to the second variable). Notice that in the present setting, condition (7.22) is equivalent to

$$\|\partial z(s,a)\| \leq 1,$$

for every $s \in I$ and $a \in E(s)$. Hence, a function $z \in \mathcal{S}^L$ is in $\mathcal{S}^{1+L}$ if and only if it is of class C^1 in a and for each $s \in I$:

(1) $z(s,0) = 0$, $\partial z(s,0) = 0$ and $z(s,E(s)) \subset F(s)$;
(2) for $a, \bar{a} \in E(s)$ we have

$$\|\partial z(s,a)\| \leq 1 \quad \text{and} \quad \|\partial z(s,a) - \partial z(s,\bar{a})\| \leq \|a - \bar{a}\|.$$

The graph of a function $z \in \mathcal{S}^{1+L}$ is given by (7.23).

Theorem 8.2 (Smooth stable manifold). *Assume that Eq. (2.12) has an exponential dichotomy on an interval I containing $\mathbb{R}_0^+$ and that there exists $\delta > 0$ such that*

$$\|\partial g(t,u)\| \leq \delta \quad \text{and} \quad \|\partial g(t,u) - \partial g(t,v)\| \leq \delta\|u - v\| \tag{8.12}$$

for every $t \in I$ and $u,v \in C$. Then, for any sufficiently small δ, the function $z \in \mathcal{S}^L$ given by Theorem 7.3 is in $\mathcal{S}^{1+L}$. Moreover, provided that

δ *is sufficiently small, for every* $t, s \in I$ *with* $t \geq s$, $a, \overline{a} \in E(s)$ *and* $j = 0, 1$
we have

$$\left\| \frac{\partial^j w_t^a}{\partial a^j} - \frac{\partial^j w_t^{\overline{a}}}{\partial a^j} \right\| \leq 3N e^{-(\lambda - 2\delta N)(t-s)} \|a - \overline{a}\|, \tag{8.13}$$

where $w_t^c = v_t(\cdot, s, c + z(s, c))$.

Proof. We consider the set $\mathfrak{X}$ of all continuous functions

$$v \colon \big\{ (t, s, a) : t \in [s - r, +\infty), s \in I, a \in E(s) \big\} \to \mathbb{R}^n$$

such that

$$\|v_t^{s,a}\| \leq 3N e^{-(\lambda - 2\delta N)(t-s)} \|a\|$$

for all $t \geq s$, where $v^{s,a} = v(\cdot, s, a)$. One can easily verify that $\mathfrak{X}$ is a
complete metric space when equipped with the norm

$$|v|_{\mathfrak{X}} := \sup\left\{ \frac{\|v_t^{s,a}\|}{\|a\|} e^{(\lambda - 2\delta N)(t-s)} : t \in [s, +\infty), s \in I, a \in E(s) \setminus \{0\} \right\}. \tag{8.14}$$

We define an operator J on $\mathfrak{X}$ by

$$(Jv)_t^{s,a} = T(t, s)a + \int_s^t T_0(t, \tau) P_0(\tau) g(\tau, v_\tau^{s,a}) \, d\tau$$
$$- \int_t^{+\infty} \overline{T}(t, \tau) Q_0(\tau) g(\tau, v_\tau^{s,a}) \, d\tau \tag{8.15}$$

for $v \in \mathfrak{X}$, $s \in I$, $a \in E(s)$ and $t \geq s$. Proceeding as in the proof of
Theorem 7.3, we find that $J(\mathfrak{X}) \subset \mathfrak{X}$ and that J is a contraction, provided
that δ is sufficiently small (note that since g is of class C^1, condition (7.2) is
equivalent to the first condition in (8.12)). Therefore, there exists a unique
$v \in \mathfrak{X}$ such that $Jv = v$. Notice that $v^{s,a} = v(\cdot, s, a)$ coincides with the
function found in the proof of Theorem 7.3.

Now we consider the set $\mathfrak{X}^{1+L}$ of all continuous functions $v \in \mathfrak{X}$ of
class C^1 in a such that

$$\|\partial v_t^{s,a}\| e^{(\lambda - 2\delta N)(t-s)} \leq 3N \tag{8.16}$$

and

$$\|\partial v_t^{s,a} - \partial v_t^{s,\overline{a}}\| e^{(\lambda - 2\delta N)(t-s)} \leq 3N \|a - \overline{a}\| \tag{8.17}$$

for each $s \in I$, $t \geq s$ and $a, \overline{a} \in E(s)$, where ∂ denotes the derivative with
respect to a. We equip the space $\mathfrak{X}^{1+L}$ with the distance in (8.14). Before
proceeding with the proof of the theorem we verify that $\mathfrak{X}^{1+L}$ is a complete

metric space. Let X and Y be Banach spaces and let $U \subset X$ be an open set. Given $c \geq 1$, we consider the set

$$D(U,Y) = \left\{ u \in C^{1+\text{Lip}}(U,Y) : |u|' \leq c \right\},$$

where $C^{1+\text{Lip}}(U,Y)$ is the set of all C^1 functions $u \colon U \to Y$ with Lipschitz derivative and

$$|u|' = |u|_\infty + \|du\|_\infty + L(du),$$

where $|\cdot|_\infty$ and $\|\cdot\|_\infty$ denote the supremum norms and

$$L(du) = \sup\left\{ \frac{\|d_x u - d_y u\|}{|x - y|} : x, y \in U \text{ and } x \neq y \right\}.$$

The following result of Henry [58] shows that $D(U,Y)$ is closed in the space of all bounded continuous functions with the supremum norm. Our proof is based on his arguments.

Lemma 8.2. *If $u_m \in D(U,Y)$ for each $m \in \mathbb{N}$ and $u \colon U \to Y$ satisfy $|u_m - u|_\infty \to 0$ when $m \to \infty$, then $u \in D(U,Y)$ and for each $x \in U$ we have $d_x u_m \to d_x u$ when $m \to \infty$.*

Proof of the lemma. Let

$$f = u_m - u_p \quad \text{and} \quad d(x) = \min\{1, \text{dist}(x, \partial U)\}.$$

For each $x \in U$ and $h \in X$ with $|h| < d(x)$, we have

$$|d_x f h| \leq |f(x + h) - f(x)| + |f(x + h) - f(x) - d_x f h|$$

$$\leq 2|f|_\infty + \left| \int_0^1 (d_{x+th} f h - d_x f h) \, dt \right|$$

$$\leq 2|f|_\infty + L(df)|h|^2.$$

Therefore,

$$\|d_x u_m - d_x u_p\| \leq \frac{2}{|h|}|u_m - u_p|_\infty + 2c|h|$$

for $|h| < d(x)$ since

$$\|d_x u_m - d_y u_m\| \leq c|x - y| \qquad (8.18)$$

for every $m \in \mathbb{N}$ and $x, y \in U$. Now we consider the function $F \colon \mathbb{R}^+ \to \mathbb{R}$ defined by

$$F(q) = \frac{2}{q}|u_m - u_p|_\infty + 2cq.$$

One can easily verify that F has a global minimum at the positive solution

$$q_0 = \sqrt{|u_m - u_p|_\infty/c}$$

of the equation $F'(q) = 0$. When $d(x) \geq q_0$, we have

$$\|d_x u_m - d_x u_p\| \leq F(q_0) = 4\sqrt{c|u_m - u_p|_\infty}.$$

On the other hand, when $d(x) \leq q_0$, we have

$$\|d_x u_m - d_x u_p\| \leq F(d(x)) = \frac{2}{d(x)}|u_m - u_p|_\infty + 2cd(x)$$

and so

$$d(x)\|d_x u_m - d_x u_p\| = 2|u_m - u_p|_\infty + 2cd(x)^2 \leq 4|u_m - u_p|_\infty.$$

Hence, provided that $|u_m - u_p|_\infty \leq 1$ we obtain

$$\|d_x u_m - d_x u_p\| \leq \frac{4}{d(x)}\sqrt{c|u_m - u_p|_\infty} \tag{8.19}$$

for all $x \in U$ (recall that $c \geq 1$).

In particular, the limit

$$v(x) = \lim_{m\to\infty} d_x u_m \tag{8.20}$$

exists for all $x \in U$, with uniform convergence on each open set $V \subset U$ away from ∂U. Hence, given $\delta > 0$, there exists $n_1 \in \mathbb{N}$ such that

$$\|d_y u_m - d_y u_p\| < \delta$$

for all $y \in V$ and $m, p > n_1$. Therefore, given $x \in V$, for all sufficiently small $h \in X$ we have

$$
\begin{aligned}
\big|u_m(x + h) &- u_m(x) - u_p(x + h) + u_p(x)\big| \\
&= \left|\int_0^1 \big(d_{x+th}u_m - d_{x+th}u_p\big)h\,dt\right| \\
&\leq \int_0^1 \|d_{x+th}u_m - d_{x+th}u_p\|\,dt|h| \leq \delta|h|
\end{aligned}
$$

for $m, p > n_1$ and so, letting $p \to \infty$, we obtain

$$|u_m(x + h) - u_m(x) - u(x + h) + u(x)| \leq \delta|h| \quad \text{for } m > n_1. \tag{8.21}$$

Now we show that u is differentiable. Note that

$$
\begin{aligned}
|u(x + h) - u(x) - v(x)h| \leq\ & |u(x + h) - u(x) - u_m(x + h) + u_m(x)| \\
& + |u_m(x + h) - u_m(x) - d_x u_m h| \\
& + |d_x u_m h - v(x)h|.
\end{aligned}
$$

$$\tag{8.22}$$

By (8.20), given $\delta > 0$, there exists $n_2 \in \mathbb{N}$ such that

$$\|d_x u_m - v(x)\| < \delta \tag{8.23}$$

for all $m > n_2$. Take an integer $m > \max\{n_1, n_2\}$. Since u_m is differentiable, we have

$$|u_m(x + h) - u_m(x) - d_x u_m h| = o(|h|). \tag{8.24}$$

Hence, it follows from (8.22) together with (8.21), (8.23) and (8.24) that

$$|u(x + h) - u(x) - v(x)h| \le 2\delta|h| + o(|h|)$$

and so

$$\limsup_{h \to 0} \frac{|u(x + h) - u(x) - v(x)h|}{|h|} \le 2\delta.$$

Since δ is arbitrary, the function u is differentiable at x and $d_x u = v(x)$. Therefore,

$$d_x u = \lim_{m \to \infty} d_x u_m$$

for all $x \in V$ and thus, for all $x \in U$. Since $u_m \in D(U, Y)$, we have $\|du_m\|_\infty \le c$ for $m \in \mathbb{N}$ and so $\|du\|_\infty \le c$. Similarly, by (8.18) we obtain

$$\|d_x u - d_y u\| \le c|x - y|$$

for every $x, y \in U$. Hence, $u \in D(U, Y)$. $\qquad\square$

It follows readily from Lemma 8.2 that $\mathcal{X}^{1+L}$ is a closed subset of $\mathcal{X}$ with the distance in (8.14).

Now let $v \in \mathcal{X}$ be the fixed point of the operator $J \colon \mathcal{X} \to \mathcal{X}$ in (8.15). We claim that $v \in \mathcal{X}^{1+L}$. Since $\mathcal{X}^{1+L}$ is a closed subset of $\mathcal{X}$, it suffices to show that

$$J(\mathcal{X}^{1+L}) \subset \mathcal{X}^{1+L}.$$

Note that the two integrals in (8.15) are of class C^1 in a. Indeed, writing

$$f(\tau, s, a) = \frac{\partial}{\partial a} g(\tau, v_\tau^{s,a}),$$

for the second integral it follows from (3.24) and (8.16) that

$$\|f(\tau, s, a)\| \le \|\partial g(\tau, v_\tau^{s,a})\| \cdot \|\partial v_\tau^{s,a}\| \le 3\delta N e^{-(\lambda - 2\delta N)(\tau - s)}$$

and

$$\|\overline{T}(t, \tau) Q_0(\tau) f(\tau, s, a)\| \le 3\delta N^2 e^{-2(\lambda - \delta N)(\tau - t)} \|a\|$$

for $\tau \geq t$, using also (8.16). Therefore, the integral

$$\int_t^{+\infty} \overline{T}(t,\tau)Q_0(\tau)f(\tau,s,a)\,d\tau$$

is well defined and converges uniformly for $a \in E(s)$. It follows from the Leibniz rule for improper integrals that the second integral in (8.15) is of class C^1 in a, with

$$\frac{\partial}{\partial a}\int_t^{+\infty} \overline{T}(t,\tau)Q_0(\tau)g(\tau,v_\tau^{s,a})\,d\tau = \int_t^{+\infty} \overline{T}(t,\tau)Q_0(\tau)f(\tau,s,a)\,d\tau. \tag{8.25}$$

A similar argument applies to the first integral in (8.15), leading to the identity

$$\partial(Jv)_s^{s,a} = T(t,s)Q(s) + \int_s^t T(t,\tau)P(\tau)f(\tau,s,a)\,d\tau$$

$$- \int_t^{+\infty} \overline{T}(t,\tau)Q_0(\tau)f(\tau,s,a)\,d\tau.$$

By Theorem 3.5 and (8.12) we obtain

$$\|\partial(Jv)_t^{s,a}\| \leq \|T(t,\tau)P(\tau)\| + \int_s^t \|T_0(t,\tau)P_0(\tau)\| \cdot \|f(\tau,s,a)\|\,d\tau$$

$$+ \int_t^{+\infty} \|\overline{T}(t,\tau)Q_0(\tau)\| \cdot \|f(\tau,s,a)\|\,d\tau$$

$$\leq Ne^{-\lambda(t-s)} + 3\delta N^2\left(\int_s^t e^{-\lambda(t-\tau)}e^{-(\lambda-2\delta N)(\tau-s)}\,d\tau\right.$$

$$\left. + \int_t^{+\infty} e^{\lambda(t-\tau)}e^{-(\lambda-2\delta N)(\tau-s)}\,d\tau\right) \tag{8.26}$$

$$\leq Ne^{-\lambda(t-s)} + 3\delta N^2\left(\int_s^t e^{-(\lambda-2\delta N)(\tau-s)}\,d\tau\right.$$

$$\left. + e^{\lambda(t-s)}\int_t^{+\infty} e^{-2(\lambda-\delta N)(\tau-s)}\,d\tau\right)$$

$$\leq Ne^{-(\lambda-2\delta N)(t-s)}(1+\theta'),$$

where

$$\theta' = 3\delta N\left(\frac{1}{\lambda-2\delta N} + \frac{1}{2(\lambda-\delta N)}\right).$$

Taking δ sufficiently small so that $\theta' \leq 2$ we have

$$\|\partial(Jv)_t^{s,a}\|e^{(\lambda-2\delta N)(t-s)} \leq 3N.$$

Since $\partial g(t,0) = 0$ for all t, using (8.12), (8.16) and (8.17), we obtain

$$
\begin{aligned}
\|f(\tau, s, a) - f(\tau, s, \overline{a})\| &\leq \|\partial g(t, v_t^{s,a}) - \partial g(t, v_t^{s,\overline{a}})\| \cdot \|\partial v_t^{s,a}\| \\
&\quad + \|\partial g(t, v_t^{s,\overline{a}})\| \cdot \|\partial v_t^{s,a} - \partial v_t^{s,\overline{a}}\| \\
&\leq 9\delta N^2 e^{-2(\lambda - 2\delta N)(t-s)} \|a - \overline{a}\| \\
&\quad + 6\delta N^2 e^{-2(\lambda - 2\delta N)(t-s)} \|a - \overline{a}\| \\
&\leq 15\delta N^2 e^{-2(\lambda - 2\delta N)(t-s)} \|a - \overline{a}\|.
\end{aligned}
$$

Hence, for each $t \geq s$ and $a, \overline{a} \in E(s)$ we have

$$
\begin{aligned}
\|\partial(Jv)_t^{s,a} - \partial(Jv)_t^{s,\overline{a}}\| &\leq \int_s^t \|T_0(t,\tau)P_0(\tau)\| \cdot \|f(\tau, s, a) - f(\tau, s, \overline{a})\| \, d\tau \\
&\quad + \int_t^{+\infty} \|\overline{T}(t,\tau)Q_0(\tau)\| \cdot \|f(\tau, s, a) - f(\tau, s, \overline{a})\| \, d\tau \\
&\leq 15\delta N^3 \|a - \overline{a}\| \left(\int_s^t e^{-\lambda(t-\tau)} e^{-2(\lambda - 2\delta N)(\tau - s)} \, d\tau \right. \\
&\quad \left. + \int_t^{+\infty} e^{\lambda(t-\tau)} e^{-2(\lambda - 2\delta N)(\tau - s)} \, d\tau \right) \\
&\leq 15\delta N^3 \|a - \overline{a}\| \left(e^{-\lambda(t-s)} \int_s^t e^{-(\lambda - 4\delta N)(\tau - s)} \, d\tau \right. \\
&\quad \left. + e^{\lambda(t-s)} \int_t^{+\infty} e^{-(3\lambda - 4\delta N)(\tau - s)} \, d\tau \right) \\
&= \theta_1 N \|a - \overline{a}\| e^{-(\lambda - 2\delta N)(t-s)}, \qquad (8.27)
\end{aligned}
$$

where

$$
\theta_1 = 15\delta N^2 \left(\frac{1}{\lambda - 4\delta N} + \frac{1}{3\lambda - 4\delta N} \right).
$$

Taking δ sufficiently small so that $\theta_1 \leq 3$ we get

$$
\|\partial(Jv)_t^{s,a} - \partial(Jv)_t^{s,\overline{a}}\| e^{(\lambda - 2\delta N)(t-s)} \leq 3N \|a - \overline{a}\|.
$$

Hence, $Jv \in \mathcal{X}^{1+L}$ and so the space $\mathcal{X}^{1+L}$ is closed in $\mathcal{X}$. Therefore, the unique fixed point of the operator J is in $\mathcal{X}^{1+L}$. In particular, in view of (8.16) and (8.17), this establishes inequality (8.13) for $j = 0, 1$.

Now we show that $z \in \mathcal{S}^{1+L}$. It follows from (7.25) and the discussion before Eq. (8.25) that z is of class C^1 in a, with

$$
\partial z(s, a) = -\int_s^{+\infty} \overline{T}(s, \tau) Q_0(\tau) f(\tau, s, a) \, d\tau.
$$

Moreover, by (8.26) (see the bounds for the second integral) we obtain

$$\|\partial z(s,a)\| \leq \frac{3\delta N^2}{2(\lambda - \delta N)} \leq 1$$

for any sufficiently small δ. Analogously, in view of (8.27) (see the bounds for the second integral),

$$\|\partial z(s,a) - \partial z(s,\overline{a})\| \leq \frac{15\delta N^3}{3\lambda - 4\delta N}\|a - \overline{a}\| \leq \|a - \overline{a}\|,$$

again for any sufficiently small δ. Hence, $z \in \mathcal{S}^{1+L}$. This concludes the proof of the theorem. $\square$

8.3   Higher Smoothness

In this section we formulate C^k stable manifolds theorems, for $k \in \mathbb{N}$, following closely the approaches in Secs. 8.1 and 8.2. The results can be obtained using induction on k in the proofs of Theorems 8.1 and 8.2 together with the approach described at length in Chap. 9 to estimate the higher derivatives. Instead of repeating all the arguments here we leave the details for the slightly more complicated setting of Chap. 9 and so we omit the proofs in the present section.

Let $I \subset \mathbb{R}$ be an interval containing $\mathbb{R}_0^+$. We assume that:

(1) $L(t): C \to \mathbb{R}^n$, for $t \in I$, are bounded linear operators such that the map $(t,v) \mapsto L(t)v$ is of class C^k and condition (2.22) holds;
(2) $g: I \times C \to \mathbb{R}^n$ is of class C^k, and $g(t,0) = 0$ and $\partial g(t,0) = 0$ for $t \in I$.

The following result is a version of class C^k of Theorem 8.1. Let $\mathcal{S}^k$ be the set of all functions $z \in \mathcal{S}^L$ (see Sec. 7.2) of class C^k in a such that $\partial z(s,0) = 0$ and

$$\|\partial^j z(s,a)\| \leq 1$$

for every $s \in I$, $a \in E(s)$ and $j = 1,\ldots,k$.

Theorem 8.3. *Assume that Eq. (2.12) has an exponential dichotomy on an interval I containing $\mathbb{R}_0^+$ and that there exists $\delta > 0$ such that*

$$\|\partial^j g(t,v)\| \leq \delta \tag{8.28}$$

for every $t \in I$, $v \in C$ and $j = 1,\ldots,k$. Then, for any sufficiently small δ, the function $z \in \mathcal{S}^L$ given by Theorem 7.3 is in $\mathcal{S}^k$.

Now we formulate a version of class C^k of Theorem 8.2. Let S^{k+L} be the set of all functions $z \in S^k$ such that

$$\|\partial^k z(s,a) - \partial^k z(s,\overline{a})\| \leq \|a - \overline{a}\|$$

for every $s \in I$ and $a, \overline{a} \in E(s)$.

Theorem 8.4. *Assume that Eq. (2.12) has an exponential dichotomy on an interval I containing $\mathbb{R}_0^+$ and that there exists $\delta > 0$ such that*

$$\|\partial^j g(t,u)\| \leq \delta \quad \text{and} \quad \|\partial^k g(t,u) - \partial^k g(t,v)\| \leq \delta \|u - v\| \qquad (8.29)$$

for every $t \in I$, $u, v \in C$ and $j = 1, \ldots, k$. Then, for any sufficiently small δ, the function $z \in S^L$ given by Theorem 7.3 is in S^{k+L}.

The proof of Theorem 8.4 uses an appropriate C^k version of Lemma 8.2. Namely, let X and Y be Banach spaces and let $U \subset X$ be an open set. Given $c \geq 1$, we consider the set

$$D^k(U, Y) = \left\{ u \in C^{k+\text{Lip}}(U, Y) : |u|'_k \leq c \right\},$$

where $C^{k+\text{Lip}}(U, Y)$ is the set of all C^k functions $u \colon U \to Y$ with Lipschitz kth derivative and

$$|u|'_k = |u|_\infty + \sum_{j=1}^{k} \|d^j u\|_\infty + L(d^k u).$$

The following lemma shows that $D^k(U, Y)$ is closed in the space of all bounded continuous functions with the supremum norm.

Lemma 8.3. *If $u_m \in D^k(U, Y)$ for each $m \in \mathbb{N}$ and $u \colon U \to Y$ satisfy $|u_m - u|_\infty \to 0$ when $m \to \infty$, then $u \in D^k(U, Y)$ and for each $x \in U$ we have $d_x^k u_m \to d_x^k u$ when $m \to \infty$.*

This result can be obtained from Lemma 8.2 using induction on k provided that we take in the induction hypothesis an arbitrary set U. The reason is that a priori the convergence of $d^k u_m$ to $d^k u$ when $m \to \infty$ need not be uniform on U, although it is uniform on each open set $V \subset U$ away from the boundary ∂U (see (8.19)). Hence, one can apply the induction hypothesis on V and then obtain the desired result writing U as a union of sets V as above.

8.4 Unstable Invariant Manifolds

In this section we formulate corresponding results to those in the former section for unstable invariant manifolds.

Let $I \subset \mathbb{R}$ be an interval containing $\mathbb{R}_0^-$ and assume that conditions 1 and 2 in Sec. 8.3 hold for some $k \in \mathbb{N}$. Moreover, let $\mathcal{U}^k$ be the set of all functions $w \in \mathcal{U}^L$ (see Sec. 7.4) of class C^k in b such that $\partial w(s, 0) = 0$ and

$$\|\partial^j w(s, b)\| \leq 1$$

for every $s \in I$, $b \in F(s)$ and $j = 1, \ldots, k$ (where ∂ denotes the partial derivative with respect to the second variable).

Theorem 8.5. *Assume that Eq. (2.12) has an exponential dichotomy on an interval I containing $\mathbb{R}_0^-$ and that there exists $\delta > 0$ such that (8.28) holds for every $t \in I$, $v \in C$ and $j = 1, \ldots, k$. Then, for any sufficiently small δ, the function $w \in \mathcal{U}^L$ given by Theorem 7.5 is in $\mathcal{U}^k$.*

Now let $\mathcal{U}^{k+L}$ be the set of all functions $w \in \mathcal{U}^k$ such that

$$\|\partial^k w(s, b) - \partial^k w(s, \overline{b})\| \leq \|b - \overline{b}\|$$

for every $s \in I$ and $b, \overline{b} \in F(s)$.

Theorem 8.6. *Assume that Eq. (2.12) has an exponential dichotomy on an interval I containing $\mathbb{R}_0^-$ and that there exists $\delta > 0$ such that (8.29) holds for every $t \in I$, $u, v \in C$ and $j = 1, \ldots, k$. Then, for any sufficiently small δ, the function $w \in \mathcal{U}^L$ given by Theorem 7.5 is in $\mathcal{U}^{k+L}$.*

PART IV
Further Topics

Chapter 9

Center Manifolds

In this chapter we construct smooth center invariant manifolds for any sufficiently small smooth perturbation of an exponential trichotomy. Namely, we construct center manifolds for C^k perturbations with Lipschitz kth derivative, with the same regularity as the perturbations. We start by introducing the notion of an exponential trichotomy, which corresponds to the existence of stable, unstable and central behaviors. In fact we allow the center direction to have some exponential behavior. In a similar manner to that for the stable and unstable manifolds, the center manifolds are characterized as the set of initial conditions for which there exists a solution with some controlled exponential speed, now both in the past and in the future.

9.1 Exponential Trichotomies

In this section we introduce a few basic notions that will be used in the construction of center invariant manifolds. First we introduce the notion of an exponential trichotomy for a linear delay equation.

Definition 9.1. We say that Eq. (2.12) has an *exponential trichotomy* (on $\mathbb{R}$) if:

(1) there exist projections $P(t), Q(t), R(t) \colon C \to C$ for $t \in \mathbb{R}$ satisfying

$$P(t) + Q(t) + R(t) = \mathrm{Id},$$

such that for $t \geq s$ we have

$$P(t)T(t,s) = T(t,s)P(s), \quad Q(t)T(t,s) = T(t,s)Q(s)$$

and

$$R(t)T(t,s) = T(t,s)R(s);$$

173

(2) the linear operator
$$\overline{T}(t,s) = T(t,s)|\operatorname{Ker} P(s)\colon \operatorname{Ker} P(s) \to \operatorname{Ker} P(t)$$
is invertible for $t \geq s$;

(3) there exist constants $\lambda, \mu, D > 0$ with $\mu < \lambda$ such that
$$\|T(t,s)R(s)\| \leq De^{\mu(t-s)}, \quad \|T(t,s)P(s)\| \leq De^{-\lambda(t-s)} \tag{9.1}$$
and
$$\|\overline{T}(s,t)R(t)\| \leq De^{\mu(t-s)}, \quad \|\overline{T}(s,t)Q(t)\| \leq De^{-\lambda(t-s)} \tag{9.2}$$
for $t \geq s$, where $\overline{T}(s,t) = \overline{T}(t,s)^{-1}$.

When Eq. (2.12) has an exponential trichotomy, the *stable*, *unstable* and *center* spaces are defined for each $t \in \mathbb{R}$ by
$$E(t) = P(t)C, \quad F(t) = Q(t)C \quad \text{and} \quad G(t) = R(t)C. \tag{9.3}$$
Clearly,
$$C = E(t) \oplus F(t) \oplus G(t) \quad \text{for } t \in \mathbb{R}.$$

Now we describe the projections of the variation of constants formula onto the stable, unstable and center directions. We start by decomposing the linear operator X_0 in (2.32). Assume that Eq. (2.12) has an exponential trichotomy. For each $t \in \mathbb{R}$ we define linear operators $P_0(t), Q_0(t), R_0(t)\colon \mathbb{R}^n \to C_0$ (with C_0 as in Sec. 2.3) by
$$P_0(t) = X_0 - Q_0(t) - R_0(t),$$
$$Q_0(t) = \overline{T}(t,t+r)Q(t+r)T(t+r,t)X_0$$
and
$$R_0(t) = \overline{T}(t,t+r)R(t+r)T(t+r,t)X_0.$$
By construction, for each $p \in \mathbb{R}^n$ we have
$$P_0(t)p \in C_0 \setminus C, \quad Q_0(t)p \in F(t) \subset C, \quad R_0(t)p \in G(t) \subset C.$$
Repeating arguments in the proof of Theorem 3.4 we obtain the following result.

Theorem 9.1. *Assume that condition (2.22) holds and that Eq. (2.12) has an exponential trichotomy. Then there exist constants $\lambda, \mu, N > 0$ with $\mu < \lambda$ such that*
$$\|T(t,s)R_0(s)\| \leq Ne^{\mu(t-s)}, \quad \|T_0(t,s)P_0(s)\| \leq Ne^{-\lambda(t-s)} \tag{9.4}$$
and
$$\|\overline{T}(s,t)R_0(t)\| \leq Ne^{\mu(t-s)}, \quad \|\overline{T}(s,t)Q_0(t)\| \leq Ne^{-\lambda(t-s)} \tag{9.5}$$
for every $t \geq s$.

We also describe the relation between the notions of an exponential trichotomy for Eq. (2.12) and of an exponential dichotomy for an arbitrary evolution family (see (2.18) for the definition).

Definition 9.2. We say that an evolution family $T(t, s)$ has an *exponential dichotomy* on an interval $I \subset \mathbb{R}$ if it satisfies the three conditions in Definition 3.2.

Given an evolution family $T(t, s)$, for each $\nu \in \mathbb{R}$ we define a new evolution family by

$$T_\nu(t, s) = e^{\nu(t-s)} T(t, s).$$

We have the following characterization of an exponential trichotomy.

Proposition 9.1. *Equation* (2.12) *has an exponential trichotomy if and only if its evolution family* $T(t, s)$ *is such that the evolution families* $T_{(\lambda+\mu)/2}(t, s)$ *and* $T_{-(\lambda+\mu)/2}(t, s)$ *have exponential dichotomies on* $\mathbb{R}$ *with projections, respectively,* $P_1(s), Q_1(s)$ *and* $P_2(s), Q_2(s)$ *satisfying*

$$E_1(s) \subset E_2(s) \quad \text{and} \quad F_2(s) \subset F_1(s) \tag{9.6}$$

for all $s \in \mathbb{R}$, *where* $E_i(s) = P_i(s)C$ *and* $F_i(s) = Q_i(s)C$ *for* $i = 1, 2$.

Proof. Assume that Eq. (2.12) has an exponential trichotomy and let $\nu = (\lambda + \mu)/2$. Clearly, the evolution family $T_\nu(t, s)$ has an exponential dichotomy on $\mathbb{R}$ with projections

$$P_1(t) = P(t) \quad \text{and} \quad Q_1(t) = Q(t) + R(t)$$

for $t \in \mathbb{R}$. Moreover,

$$\|T_\nu(t, s)P_1(s)\| \le Ne^{(-\lambda+\nu)(t-s)}, \quad \|T_\nu(s, t)Q_1(t)\| \le 2Ne^{(-\lambda+\nu)(t-s)}$$

for $t \ge s$. Similarly, the evolution family $T_{-\nu}(t, s)$ has an exponential dichotomy on $\mathbb{R}$ with projections

$$P_2(t) = P(t) + R(t) \quad \text{and} \quad Q_2(t) = Q(t)$$

for $t \in \mathbb{R}$. Moreover,

$$\|T_{-\nu}(t, s)P_2(s)\| \le 2Ne^{(-\lambda+\nu)(t-s)}, \quad \|T_{-\nu}(s, t)Q_2(t)\| \le Ne^{(-\lambda+\nu)(t-s)}$$

for $t \ge s$.

We want to show that (9.6) holds for every $s \in \mathbb{R}$. Let

$$\Lambda(\varphi) = \limsup_{t \to +\infty} \frac{1}{t} \log \|T_\nu(t, s)\varphi\|.$$

We proceed by contradiction. Assume that there exists $\varphi \in E_1(s) \setminus E_2(s)$ and write $\varphi = z + w$, with $z \in E_2(s)$ and $w \in F_2(s) \setminus \{0\}$. Since $\varphi \in E_1(s)$, we have

$$\|T_\nu(t,s)\varphi\| \leq N e^{(-\lambda+\nu)(t-s)} \|\varphi\|$$

and so $\Lambda(\varphi) \leq -\lambda + \nu$. On the other hand, since $w \neq 0$ we have

$$\Lambda(\varphi) = \max\{\Lambda(z), \Lambda(w)\} = \Lambda(w) = \limsup_{t \to +\infty} \frac{1}{t} \log\|T_\nu(t,s)w\|$$

and it follows from

$$\|T_\nu(t,s)w\| = e^{2\nu(t-s)}\|T_{-\nu}(t,s)w\| \geq \frac{1}{N} e^{(\lambda+\nu)(t-s)} \|w\|$$

that $\Lambda(\varphi) \geq \lambda + \nu$. But this contradicts the inequality $\Lambda(\varphi) \leq -\lambda + \nu$. Therefore, $E_1(s) \subset E_2(s)$. One can also show that $F_2(s) \subset F_1(s)$ for $s \in \mathbb{R}$.

Now we prove the converse. Assume that the evolution families $T_\nu(t,s)$ and $T_{-\nu}(t,s)$ have exponential dichotomies on $\mathbb{R}$ with stable and unstable spaces satisfying (9.6). First we show that

$$(E_2(s) \cap F_1(s)) \oplus E_1(s) \oplus F_2(s) = C \qquad (9.7)$$

for every $s \in \mathbb{R}$. It follows from $E_1(s) \oplus F_1(s) = C$ that

$$(E_2(s) \cap E_1(s)) \oplus (E_2(s) \cap F_1(s)) = E_2(s).$$

In view of (9.6) we have $E_2(s) \cap E_1(s) = E_1(s)$ and hence,

$$E_1(s) \oplus (E_2(s) \cap F_1(s)) = E_2(s).$$

Identity (9.7) follows now from the fact that $E_2(s) \oplus F_2(s) = C$. Moreover, we show that

$$P_1(s)Q_2(s) = Q_2(s)P_1(s) = 0 \qquad (9.8)$$

for each $s \in \mathbb{R}$. By (9.6), for each $\varphi \in C$ we have $Q_2(s)\varphi \in F_2(s) \subset F_1(s)$ and hence,

$$P_1(s)Q_2(s)\varphi \in P_1(s)F_1(s) = \{0\}.$$

Similarly, again by (9.6), for each $\varphi \in C$ we have $P_1(s)\varphi \in E_1(s) \subset E_2(s)$ and hence,

$$Q_2(s)P_1(s)\varphi \in Q_2(s)E_2(s) = \{0\},$$

which establishes (9.8). We also want to identify explicitly the projections associated with the decomposition in (9.7). Let

$$P(t) = P_1(t), \quad Q(t) = Q_2(t) \quad \text{and} \quad R(t) = \text{Id} - P_1(t) - Q_2(t). \qquad (9.9)$$

Since $T_\nu(t,s)$ and $T_{-\nu}(t,s)$ have exponential dichotomies on $\mathbb{R}$, we obtain

$$T_\nu(t,s)P(s) = P(t)T_\nu(t,s), \quad T_{-\nu}(t,s)Q(s) = Q(t)T_{-\nu}(t,s).$$

Moreover, since

$$T_\nu(t,s) = e^{\nu(t-s)}T(t,s) \quad \text{and} \quad T_{-\nu}(t,s) = e^{-\nu(t-s)}T(t,s),$$

we conclude that

$$T(t,s)P(s) = P(t)T(t,s), \quad T(t,s)Q(s) = Q(t)T(t,s).$$

This readily implies that

$$T(t,s)R(s) = R(t)T(t,s).$$

In order to show that Eq. (2.12) has an exponential trichotomy, it remains to prove that the operator $R(t)$ in (9.9) is a projection and that we have the appropriate exponential bounds. By (9.8) we have

$$\begin{aligned}
R(t)^2 &= (\mathrm{Id} - P_1(t) - Q_2(t))^2 \\
&= \mathrm{Id} - 2P_1(t) - 2Q_2(t) + P_1(t)^2 + Q_2(t)^2 + P_1(t)Q_2(t) + Q_2(t)P_1(t) \\
&= \mathrm{Id} - P_1(t) - Q_2(t) = R(t)
\end{aligned}$$

and so the operator $R(t)$ is a projection. Now we consider the spaces in (9.3). We claim that

$$E(t) = E_1(t), \quad F(t) = F_2(t), \quad G(t) = E_2(t) \cap F_1(t). \tag{9.10}$$

The first two identities are clear. For the third identity, note that by (9.8) we have

$$P(t)R(t) = Q(t)R(t) = 0$$

and hence,

$$R(t)C \subset F_1(t) \quad \text{and} \quad R(t)C \subset E_2(t).$$

Thus,

$$R(t)C \subset E_2(t) \cap F_1(t)$$

and so the third identity in (9.10) follows readily from (9.7). Finally, we establish the inequalities in (9.1) and (9.2). Note that

$$\begin{aligned}
\|T(t,s)P(s)\| &= \|T_\nu(t,s)e^{-\nu(t-s)}P_1(s)\| \\
&\leq Ne^{-\nu(t-s)}e^{(-\lambda+\nu)(t-s)} = Ne^{-\lambda(t-s)}
\end{aligned}$$

and

$$\begin{aligned}
\|\overline{T}(s,t)Q(t)\| &= \|\overline{T}_{-\nu}(s,t)e^{-\nu(t-s)}Q_2(t)\| \\
&\leq Ne^{-\nu(t-s)}e^{(-\lambda+\nu)(t-s)} = Ne^{-\lambda(t-s)}
\end{aligned}$$

for $t \geq s$. Moreover,

$$
\begin{aligned}
\|T(t,s)R(s)\| &\leq \|T(t,s)|(E_2(s) \cap F_1(s))\| \cdot \|R(s)\| \\
&\leq \|T(t,s)|E_2(s)\| \cdot \|R(s)\| \\
&= e^{\nu(t-s)}\|T_{-\nu}(t,s)|E_2(s)\| \cdot \|R(s)\|
\end{aligned} \tag{9.11}
$$

and

$$
\begin{aligned}
\|\overline{T}(s,t)R(t)\| &\leq \|\overline{T}(s,t)|F_1(t)\| \cdot \|R(t)\| \\
&= e^{\nu(t-s)}\|\overline{T}_\nu(s,t)|F_1(t)\| \cdot \|R(t)\|
\end{aligned} \tag{9.12}
$$

for $t \geq s$. Since

$$
\|R(t)\| = \|\mathrm{Id} - P_1(t) - Q_2(t)\| \leq 1 + 2N,
$$

it follows, respectively, from (9.11) and (9.12) that

$$
\|T(t,s)R(s)\| \leq (1+2N)2Ne^{\nu(t-s)}e^{(-\lambda+\nu)(t-s)} = (1+2N)2Ne^{\mu(t-s)}
$$

and

$$
\|\overline{T}(s,t)R(t)\| \leq (1+2N)2Ne^{\nu(t-s)}e^{(-\lambda+\nu)(t-s)} = (1+2N)2Ne^{\mu(t-s)}.
$$

This shows that Eq. (2.12) has an exponential trichotomy. $\qquad\square$

9.2   Properties of the Center Set

In this section we introduce the notion of center set for Eq. (7.1) and we describe some of its basic properties. More precisely, we consider the perturbations of Eq. (2.12) of the form (7.1), where:

(1) $L(t) \colon C \to \mathbb{R}^n$, for $t \in \mathbb{R}$, are bounded linear operators such that the map $(t,v) \mapsto L(t)v$ is continuous on $\mathbb{R} \times C$ and condition (2.22) holds;
(2) $g \colon \mathbb{R} \times C \to \mathbb{R}^n$ is continuous and $g(t,0) = 0$ for all $t \in \mathbb{R}$.

We also assume that g is Lipschitz in the second variable. Given $(s,\varphi) \in \mathbb{R} \times C$, we continue to denote by $v(\cdot,s,\varphi)$ the unique solution of Eq. (7.1) on $[s-r,+\infty)$ with $v_s = \varphi$ (all the solutions considered in this section can in fact be continued to the whole $\mathbb{R}$).

Let $\mathcal{V}_\lambda$ be the set of all continuous functions $v \colon \mathbb{R} \to \mathbb{R}^n$ such that

$$
\sup_{t \in \mathbb{R}} \left(\|v_t\| e^{-\nu|t|} \right) < +\infty
$$

for any sufficiently large $\nu < \lambda$.

Definition 9.3. The *center set* V^c of Eq. (7.1) is the set of all initial conditions $(s,\varphi) \in \mathbb{R} \times C$ for which there exists a solution $v \in \mathcal{V}_\lambda$ of the equation with $v_s = \varphi$.

The center set has the following invariance property, which can be established in a similar manner to that in the proof of Proposition 7.1.

Proposition 9.2. *If $(s, \varphi) \in V^c$, then $(t, v_t) \in V^c$ for all $t \in \mathbb{R}$, where $v \in \mathcal{V}_\lambda$ is the solution in Definition 9.3.*

We start with a characterization of the center set.

Proposition 9.3. *If Eq. (2.12) has an exponential trichotomy, then its center set is composed of the pairs $(s, \varphi) \in \mathbb{R} \times C$ for which there exists a function $v : \mathbb{R} \to \mathbb{R}^n$ in $\mathcal{V}_\lambda$ such that $v_s = \varphi$,*

$$
v_t = T(t, s)R(s)\varphi + \int_s^t T(t, \tau)R_0(\tau)g(\tau, v_\tau)\, d\tau
$$
$$
+ \int_{-\infty}^t T_0(t, \tau)P_0(\tau)g(\tau, v_\tau)\, d\tau - \int_t^{+\infty} \overline{T}(t, \tau)Q_0(\tau)g(\tau, v_\tau)\, d\tau
\tag{9.13}
$$

for $t \geq s$ and

$$
v_t = \overline{T}(t, s)R(s)\varphi - \int_t^s \overline{T}(t, \tau)R_0(\tau)g(\tau, v_\tau)\, d\tau
$$
$$
+ \int_{-\infty}^t T_0(t, \tau)P_0(\tau)g(\tau, v_\tau)\, d\tau - \int_t^{+\infty} \overline{T}(t, \tau)Q_0(\tau)g(\tau, v_\tau)\, d\tau
\tag{9.14}
$$

for $t \leq s$.

Proof. Let v be a solution of Eq. (7.1) in $\mathcal{V}_\lambda$ with $v_s = \varphi$. By Theorem 3.6, we have

$$
R(\tau)v_\tau = T(\tau, \sigma)R(\sigma)v_\sigma + \int_\sigma^\tau T_0(\tau, u)R_0(u)g(u, v_u)\, du,
$$
$$
P(\tau)v_\tau = T(\tau, \sigma)P(\sigma)v_\sigma + \int_\sigma^\tau T_0(\tau, u)P_0(u)g(u, v_u)\, du,
\tag{9.15}
$$
$$
Q(\tau)v_\tau = T(\tau, \sigma)Q(\sigma)v_\sigma + \int_\sigma^\tau T(\tau, u)Q_0(u)g(u, v_u)\, du
$$

for $\tau \geq \sigma$. The third identity also holds for $\tau \leq \sigma$, with T replaced by $\overline{T}$. Therefore,

$$
Q(t)v_t = \overline{T}(t, \sigma)Q(\sigma)v_\sigma - \int_t^\sigma \overline{T}(t, u)Q_0(u)g(u, v_u)\, du
\tag{9.16}
$$

for $t \leq \sigma$. Since v belongs to $\mathcal{V}_\lambda$, by (9.2) and (9.5) we obtain

$$
\|\overline{T}(t, \sigma)Q(\sigma)v_\sigma\| \leq N e^{-\lambda(\sigma - t)} e^{\nu|\sigma|} \sup_{u \in \mathbb{R}} \left(\|v_u\| e^{-\nu|u|} \right) \to 0
$$

when $\sigma \to +\infty$. Thus,

$$Q(t)v_t = -\int_t^{+\infty} \overline{T}(t,u)Q_0(u)g(u,v_u)\,du. \tag{9.17}$$

Similarly, taking $\tau = t$, it follows from the second identity in (9.15) that

$$P(t)v_t = T(t,\sigma)P(\sigma)v_\sigma + \int_\sigma^t T_0(t,u)P_0(u)g(u,v_u)\,du.$$

Since v belongs to $\mathcal{V}_\lambda$, by (9.1) and (9.4) we obtain

$$\|\overline{T}(t,\sigma)P(\sigma)v_\sigma\| \le Ne^{-\lambda(t-\sigma)}e^{\nu|\sigma|}\sup_{u\in\mathbb{R}}\big(\|v_u\|e^{-\nu|u|}\big) \to 0$$

when $\sigma \to -\infty$. Hence,

$$P(t)v_t = \int_{-\infty}^t T_0(t,u)P_0(u)g(u,v_u)\,du. \tag{9.18}$$

Adding the first identity in (9.15) with $\tau = t$ and $\sigma = s$ to (9.17) and (9.18) yields identity (9.13). Similarly, adding (9.16) with $\sigma = s$ to (9.17) and (9.18) yields identity (9.14).

Now assume that (9.13) holds for all $t \ge s$ and some function $v\colon \mathbb{R} \to \mathbb{R}^n$ in $\mathcal{V}_\lambda$ with $v_s = \varphi$. Taking $t = s$, we obtain

$$v_s = R(s)v_s + \int_{-\infty}^s T_0(s,\tau)P_0(\tau)g(\tau,v_\tau)\,d\tau - \int_s^{+\infty} \overline{T}(s,\tau)Q_0(\tau)g(\tau,v_\tau)\,d\tau$$

and so

$$P(s)v_s = \int_{-\infty}^s T_0(s,\tau)P_0(\tau)g(\tau,v_\tau)\,d\tau,$$

$$Q(s)v_s = -\int_s^{+\infty} \overline{T}(s,\tau)Q_0(\tau)g(\tau,v_\tau)\,d\tau.$$

This implies that for each $t \ge s$ we have

$$T(t,s)v_s + \int_s^t T_0(t,\tau)X_0 g(\tau,v_\tau)\,d\tau$$

$$= T(t,s)R(s)v_s + \int_s^t T_0(t,\tau)R_0(\tau)g(\tau,v_\tau)\,d\tau$$

$$+ \int_s^t T(t,\tau)P_0(\tau)g(\tau,v_\tau)\,d\tau + \int_s^t T(t,\tau)Q_0(\tau)g(\tau,v_\tau)\,d\tau$$

$$+ T(t,s)P(s)v_s + T(t,s)Q(s)v_s$$

$$= R(t)v_t + \int_s^t T(t,\tau)P_0(\tau)g(\tau, v_\tau)\, d\tau$$

$$+ T(t,s)\int_{-\infty}^s T_0(s,\tau)P_0(\tau)g(\tau, v_\tau)\, d\tau + \int_s^t T(t,\tau)Q_0(\tau)g(\tau, v_\tau)\, d\tau$$

$$- T(t,s)\int_s^{+\infty} \overline{T}(s,\tau)Q_0(\tau)g(\tau, v_\tau)\, d\tau$$

$$= R(t)v_t + \int_{-\infty}^t T(t,\tau)P_0(\tau)g(\tau, v_\tau)\, d\tau - \int_t^{+\infty} \overline{T}(t,\tau)Q_0(\tau)g(\tau, v_\tau)\, d\tau$$

$$= R(t)v_t + P(t)v_t + Q(t)v_t = v_t.$$

This shows that v is a solution of Eq. (7.1) on $[s - r, +\infty)$. Similarly, one can show that if (9.14) holds for all $t \le s$ and some function $v: \mathbb{R} \to \mathbb{R}^n$ in $\mathcal{V}_\lambda$ with $v_s = \varphi$, then v is a solution of Eq. (7.1) on $(-\infty, s]$. This completes the proof of the proposition. $\qquad\square$

Using the characterization of the center set given by Proposition 9.3, one can show that if Eq. (2.12) has an exponential trichotomy, then all solutions in $\mathcal{V}_\lambda$ satisfy a certain upper bound.

Proposition 9.4. *If Eq. (2.12) has an exponential trichotomy and there exists $\delta > 0$ such that (7.2) holds for every $t \in \mathbb{R}$ and $u, v \in C$, then, for any sufficiently small δ, each solution $v: \mathbb{R} \to \mathbb{R}^n$ of Eq. (7.1) in $\mathcal{V}_\lambda$ with $v_s = \varphi$ satisfies*

$$\|v_t\| \le 2N e^{(\mu + 4\delta N)|t-s|}\|R(s)\varphi\| \tag{9.19}$$

for every $t, s \in \mathbb{R}$.

Proof. By Proposition 9.3, each solution $v \in \mathcal{V}_\lambda$ of Eq. (7.1) with $v_s = \varphi$ satisfies (9.13) and (9.14). On the other and, by (7.2) together with the fact that $g(t, 0) = 0$, it follows from the identities (9.13) and (9.14) that

$$\|v_t\| \le N e^{\mu(t-s)}\|R(s)\varphi\| + \delta N \int_s^t e^{\mu(t-\tau)}\|v_\tau\|\, d\tau$$

$$+ \delta N \int_{-\infty}^t e^{-\lambda(t-\tau)}\|v_\tau\|\, d\tau + \delta N \int_t^{+\infty} e^{\lambda(t-\tau)}\|v_\tau\|\, d\tau$$

for $t \ge s$ and

$$\|v_t\| \le N e^{\mu(s-t)}\|R(s)\varphi\| + \delta N \int_t^s e^{\mu(\tau-t)}\|v_\tau\|\, d\tau$$

$$+ \delta N \int_{-\infty}^t e^{-\lambda(t-\tau)}\|v_\tau\|\, d\tau + \delta N \int_t^{+\infty} e^{\lambda(t-\tau)}\|v_\tau\|\, d\tau$$

for $t \leq s$. Since $v \in \mathcal{V}_\lambda$, the function

$$\alpha(t) = \|v_t\| e^{-\nu|t-s|}$$

is bounded on $\mathbb{R}$. Furthermore, it satisfies

$$\alpha(t) \leq N e^{-(\nu-\mu)(t-s)} \|R(s)\varphi\| + \delta N \int_s^t e^{-(\nu-\mu)(t-\tau)} \alpha(\tau) \, d\tau$$

$$+ \delta N \int_{-\infty}^t e^{-(\lambda+\nu)(t-\tau)} \alpha(\tau) \, d\tau + \delta N \int_t^{+\infty} e^{(\lambda-\nu)(t-\tau)} \alpha(\tau) \, d\tau$$

$$\leq N e^{-(\nu-\mu)(t-s)} \|R(s)\varphi\| + \delta N \int_s^t e^{-(\nu-\mu)(t-\tau)} \alpha(\tau) \, d\tau$$

$$+ \delta N \int_{-\infty}^t e^{-(\nu-\mu)(t-\tau)} \alpha(\tau) \, d\tau + \delta N \int_t^{+\infty} e^{(\nu-\mu)(t-\tau)} \alpha(\tau) \, d\tau$$

$$\tag{9.20}$$

for $t \geq s$ and

$$\alpha(t) \leq N e^{-(\nu-\mu)(s-t)} \|R(s)\varphi\| + \delta N \int_t^s e^{-(\nu-\mu)(\tau-t)} \alpha(\tau) \, d\tau$$

$$+ \delta N \int_{-\infty}^t e^{-(\nu-\mu)(t-\tau)} \alpha(\tau) \, d\tau + \delta N \int_t^{+\infty} e^{(\nu-\mu)(t-\tau)} \alpha(\tau) \, d\tau$$

$$\tag{9.21}$$

for $t \leq s$.

Before proceeding with the proof, we establish an auxiliary result.

Lemma 9.1. *Given $s \in \mathbb{R}$, let $x \colon \mathbb{R} \to \mathbb{R}_0^+$ be a bounded continuous function such that*

$$x(t) \leq \beta e^{-\gamma(t-s)} + \delta D \int_s^t e^{-\gamma(t-\tau)} x(\tau) \, d\tau + \delta D \int_{-\infty}^t e^{-\gamma(t-\tau)} x(\tau) \, d\tau$$

$$+ \delta D \int_t^{+\infty} e^{\gamma(t-\tau)} x(\tau) \, d\tau$$

$$\tag{9.22}$$

for $t \geq s$ and

$$x(t) \leq \beta e^{-\gamma(s-t)} + \delta D \int_t^s e^{-\gamma(\tau-t)} x(\tau) \, d\tau + \delta D \int_{-\infty}^t e^{-\gamma(t-\tau)} x(\tau) \, d\tau$$

$$+ \delta D \int_t^{+\infty} e^{\gamma(t-\tau)} x(\tau) \, d\tau$$

$$\tag{9.23}$$

for $t \leq s$, with $\gamma > 0$. Then, for any sufficiently small δ, we have

$$x(t) \leq 2\beta e^{-(\gamma-4\delta N)|t-s|} \quad \text{for } t \in \mathbb{R}. \tag{9.24}$$

Proof of the lemma. The argument is an elaboration of the proof of Lemma 7.1. We first show that $x(t) \le y(t)$ for $t \in \mathbb{R}$, where y is the unique bounded continuous function satisfying

$$y(t) = \beta e^{-\gamma(t-s)} + \delta D \int_s^t e^{-\gamma(t-\tau)} y(\tau)\, d\tau + \delta D \int_{-\infty}^t e^{-\gamma(t-\tau)} y(\tau)\, d\tau$$
$$+ \delta D \int_t^{+\infty} e^{\gamma(t-\tau)} y(\tau)\, d\tau$$

$$\tag{9.25}$$

for $t \ge s$ and

$$y(t) = \beta e^{-\gamma(s-t)} + \delta D \int_t^s e^{-\gamma(\tau-t)} y(\tau)\, d\tau + \delta D \int_{-\infty}^t e^{-\gamma(t-\tau)} y(\tau)\, d\tau$$
$$+ \delta D \int_t^{+\infty} e^{\gamma(t-\tau)} y(\tau)\, d\tau$$

$$\tag{9.26}$$

for $t \le s$. One can easily verify that

$$y'' = \delta D y' + (\gamma^2 - 3\delta D\gamma)y$$

for $t \ge s$. Since y is bounded, we have

$$y(t) = y(s)e^{-\overline{\gamma}(t-s)}, \quad \text{where } \overline{\gamma} = \frac{1}{2}\left(\sqrt{4(\gamma^2 - 3\delta D\gamma) + \delta^2 D^2} - \delta D\right).$$

Substituting $y(t)$ in (9.25) and taking $t = s$, we find that

$$y(s) = \beta + \delta D y(s) \int_{-\infty}^s e^{-(\gamma-\overline{\gamma})(s-\tau)}\, d\tau + \delta D y(s) \int_s^{+\infty} e^{(\gamma+\overline{\gamma})(s-\tau)}\, d\tau$$

$$= \beta + \delta D y(s)\left(\frac{1}{\gamma-\overline{\gamma}} + \frac{1}{\gamma+\overline{\gamma}}\right) = \beta + \frac{2\gamma\delta D}{\gamma^2 - \overline{\gamma}^2} y(s)$$

$$= \beta + \frac{2\gamma}{3\gamma + \overline{\gamma}} y(s)$$

for any sufficiently small δ. Therefore,

$$y(s) = \frac{\beta(3\gamma + \overline{\gamma})}{\gamma + \overline{\gamma}} \tag{9.27}$$

and so

$$y(t) = \frac{\beta(3\gamma + \overline{\gamma})}{\gamma + \overline{\gamma}} e^{-\overline{\gamma}(t-s)} \quad \text{for } t \ge s.$$

Similarly, one can easily verify that

$$y'' = -\delta D y' + (\gamma^2 - 3\delta D\gamma)y$$

for $t \le s$. Since y is bounded for $t \le s$, we have

$$y(t) = y(s)e^{\overline{\gamma}(t-s)} \quad \text{for } t \le s.$$

Substituting $y(t)$ in (9.26) and taking $t = s$, we find that (9.27) holds for any sufficiently small δ. Therefore,

$$y(t) = \frac{\beta(3\gamma + \overline{\gamma})}{\gamma + \overline{\gamma}}e^{-\overline{\gamma}|t-s|} \quad \text{for } t \in \mathbb{R}.$$

Now we consider the function

$$z(t) = x(t) - y(t) \quad \text{for } t \in \mathbb{R}.$$

By (9.22), (9.23), (9.25) and (9.26) we obtain

$$
z(t) \le \delta D \left| \int_s^t e^{-\gamma|t-\tau|} z(\tau)\, d\tau \right| + \delta D \int_{-\infty}^t e^{-\gamma(t-\tau)} z(\tau)\, d\tau \\
+ \delta D \int_t^{+\infty} e^{\gamma(t-\tau)} z(\tau)\, d\tau
\tag{9.28}
$$

for $t \in \mathbb{R}$. Since x and y are bounded, we have $z := \sup_{t \in \mathbb{R}} z(t) < +\infty$ and it follows from (9.28) that

$$
z \le \delta D z \sup_{t \in \mathbb{R}} \left| \int_s^t e^{-\gamma|t-\tau|}\, d\tau \right| + \delta D z \sup_{t \in \mathbb{R}} \int_{-\infty}^t e^{-\gamma(t-\tau)}\, d\tau \\
+ \delta D z \sup_{t \in \mathbb{R}} \int_t^{+\infty} e^{\gamma(t-\tau)}\, d\tau \le \delta D z \frac{3}{\gamma}.
$$

Taking δ sufficiently small, we obtain $z \le 0$ and so $z(t) \le 0$ for all $t \in \mathbb{R}$. Therefore,

$$x(t) \le y(t) = \frac{\beta(3\gamma + \overline{\gamma})}{\gamma + \overline{\gamma}}e^{-\overline{\gamma}|t-s|}. \tag{9.29}$$

Taking δ sufficiently small, we obtain $\overline{\gamma} \ge \gamma - 4\delta N$ and

$$\frac{\beta(3\gamma - \overline{\gamma})}{\gamma + \overline{\gamma}} = \beta\left(1 + \frac{2\gamma}{\gamma + \overline{\gamma}}\right) \le \beta\left(1 + \frac{2\gamma}{2\gamma - 4\delta N}\right) \le 2\beta.$$

By (9.29), we conclude that property (9.24) holds.   □

By (9.20) and (9.21), it follows from Lemma 9.1 with

$$x(t) = \alpha(t), \quad \beta = N\|R(s)\varphi\|, \quad \gamma = \nu - \mu > 0 \quad \text{and} \quad D = N$$

that

$$\alpha(t) \le 2Ne^{-(\nu-\mu-4\delta N)|t-s|}\|R(s)\varphi\|.$$

Therefore,

$$\|v_t\| = \alpha(t)e^{\nu|t-s|} \le 2Ne^{(\mu+4\delta N)|t-s|}\|R(s)\varphi\|,$$

for $t \in \mathbb{R}$, which establishes (9.19).   □

9.3 Lipschitz Center Manifolds

In this section we show that there exist Lipschitz center invariant manifolds for Eq. (7.1). We shall always assume that conditions 1 and 2 in Sec. 9.2 hold and that Eq. (2.12) has an exponential trichotomy.

Let $\mathcal{C}^L$ be the set of all continuous functions

$$z\colon \big\{(s,c) \in \mathbb{R} \times C : c \in G(s)\big\} \to C$$

such that for each $s \in \mathbb{R}$:

(1) $z(s,0) = 0$ and $z(s, G(s)) \subset E(s) \oplus F(s)$;
(2) for $c, \bar{c} \in G(s)$ we have

$$\|z(s,c) - z(s,\bar{c})\| \le \|c - \bar{c}\|.$$

For each $z \in \mathcal{C}^L$ we consider its graph

$$\text{graph } z = \big\{(s, c + z(s,c)) : (s,c) \in \mathbb{R} \times G(s)\big\} \subset \mathbb{R} \times C. \tag{9.30}$$

The following result is a Lipschitz center manifold theorem.

Theorem 9.2 (Lipschitz center manifold). *If Eq. (2.12) has an exponential trichotomy and there exists $\delta > 0$ such that (7.2) holds for every $t \in \mathbb{R}$ and $u, v \in C$, then, for any sufficiently small δ, there exists a function $z \in \mathcal{C}^L$ such that $V^c = \text{graph } z$. Moreover, for each initial condition $(s, c + z(s,c)) \in V^c$ there exists a unique solution $v = v^{s,c}$ of Eq. (7.1) in V_λ with $v_s = c + z(s,c)$, which satisfies*

$$\big\|v_t^{s,c} - v_t^{s,\bar{c}}\big\| \le 2N e^{(\mu + 4\delta N)|t-s|} \|c - \bar{c}\| \tag{9.31}$$

for every $t, s \in \mathbb{R}$ and $c, \bar{c} \in G(s)$.

Proof. Given $s \in \mathbb{R}$ and $c \in G(s)$, let $\mathcal{Z}_+^c$ be the set of all continuous functions $v\colon [s - r, +\infty) \to \mathbb{R}^n$ with $R(s)v_s = c$ such that

$$\|v_t\| \le 2N e^{(\mu + 4\delta N)(t-s)} \|c\|$$

for $t \ge s$ and let $\mathcal{Z}_-^c$ be the set of all continuous functions $v\colon (-\infty, s] \to \mathbb{R}^n$ with $R(s)v_s = c$ such that

$$\|v_t\| \le 2N e^{(\mu + 4\delta N)(s-t)} \|c\|$$

for $t \le s$. Clearly, $\mathcal{Z}_+^c$ and $\mathcal{Z}_-^c$ are complete metric spaces when equipped, respectively, with the norms

$$|v|_{\mathcal{Z}_+^c} := \sup\big\{\|v_t\| e^{-(\mu + 4\delta N)(t-s)} : t \ge s\big\}$$

and

$$|v|_{Z^c_-} := \sup\{\|v_t\|e^{-(\mu+4\delta N)(s-t)} : t \le s\}.$$

We define an operator J on Z^c_+ by

$$(Jv)_t = T(t,s)c + \int_s^t T(t,\tau)R_0(\tau)g(\tau,v_\tau)\,d\tau$$

$$+ \int_{-\infty}^t T_0(t,\tau)P_0(\tau)g(\tau,v_\tau)\,d\tau - \int_t^{+\infty} \overline{T}(t,\tau)Q_0(\tau)g(\tau,v_\tau)\,d\tau$$

for each $v \in Z^c_+$ and $t \ge s$. Note that the function Jv is continuous. Moreover,

$$(Jv)_s = c + \int_{-\infty}^s T_0(s,\tau)P_0(\tau)g(\tau,v_\tau)\,d\tau - \int_s^{+\infty} \overline{T}(s,\tau)Q_0(\tau)g(\tau,v_\tau)\,d\tau$$

and so $R(s)(Jv)_s = c$. Now we show that $J(Z^c_+) \subset Z^c_+$ and that J is a contraction. For $t \ge s$ and $v, w \in Z^c_+$, by (7.2), (9.4) and (9.5) we have

$$\|(Jv)_t - (Jw)_t\| \le N\delta \int_s^t e^{\mu(t-\tau)}\|v_\tau - w_\tau\|\,d\tau$$

$$+ N\delta \int_{-\infty}^t e^{-\lambda(t-\tau)}\|v_\tau - w_\tau\|\,d\tau$$

$$+ N\delta \int_t^{+\infty} e^{\lambda(t-\tau)}\|v_\tau - w_\tau\|\,d\tau$$

$$\le N\delta e^{(\mu+4\delta N)(t-s)} \int_s^t e^{-4\delta N(t-\tau)}\,d\tau |v-w|_{Z^c_+}$$

$$+ N\delta e^{(\mu+4\delta N)(t-s)} \int_{-\infty}^t e^{-(\lambda+\mu+4\delta N)(t-\tau)}\,d\tau |v-w|_{Z^c_+}$$

$$+ N\delta e^{\lambda t - \mu s - 4\delta N s} \int_t^{+\infty} e^{-(\lambda-\mu-4\delta N)\tau}\,d\tau |v-w|_{Z^c_+}$$

$$\le \theta e^{(\mu+4\delta N)(t-s)}|v-w|_{Z^c_+},$$

where

$$\theta = \frac{1}{4} + \frac{\delta N}{\lambda + \mu + 4\delta N} + \frac{\delta N}{\lambda - \mu - 4\delta N}.$$

Therefore,

$$|Jv - Jw|_{Z^c_+} \le \theta |v-w|_{Z^c_+}.$$

For any sufficiently small δ such that $\theta < 1/2$, the operator J is a contraction. Moreover, by (9.1), we have $|J0|_{Z_+^c} \leq N\|c\|$ and thus,

$$
\begin{aligned}
|Jv|_{Z_+^c} &\leq |J0|_{Z_+^c} + |Jv - J0|_{Z_+^c} \\
&\leq N\|c\| + \theta|v|_{Z_+^c} \\
&\leq N\|c\| + \frac{1}{2}2N\|c\| = 2N\|c\|.
\end{aligned}
$$

Hence, $J(Z_+^c) \subset Z_+^c$ and there exists a unique $v = v_+^{s,c} \in Z_+^c$ such that $Jv = v$. Similarly, there exists a unique $v = v_-^{s,c} \in Z_-^c$ satisfying

$$
\begin{aligned}
v_t = \overline{T}(t,s)R(s)\varphi &- \int_t^s \overline{T}(t,\tau)R_0(\tau)g(\tau,v_\tau)\,d\tau \\
&+ \int_{-\infty}^t T_0(t,\tau)P_0(\tau)g(\tau,v_\tau)\,d\tau - \int_t^{+\infty} \overline{T}(t,\tau)Q_0(\tau)g(\tau,v_\tau)\,d\tau
\end{aligned} \tag{9.32}
$$

for $t \leq s$. Finally, we define a function $v^{s,c}\colon \mathbb{R} \to \mathbb{R}^n$ by

$$
v^{s,c}(t) = \begin{cases} v_+^{s,c}(t) & \text{if } t \geq s, \\ v_-^{s,c}(t) & \text{if } t < s. \end{cases} \tag{9.33}
$$

Since

$$
(v_+^{s,c})_s = (v_-^{s,c})_s \quad \text{for } s \in \mathbb{R}, c \in G(s),
$$

the function $v^{s,c}$ is continuous and satisfies

$$
\|v_t^{s,c}\| \leq 2Ne^{(\mu+4\delta N)|t-s|}\|c\| \quad \text{for } t \in \mathbb{R}.
$$

Now we define a map

$$
z\colon \{(s,c) \in \mathbb{R} \times C \,:\, c \in G(s)\} \to C
$$

by

$$
\begin{aligned}
z(s,c) &= \int_{-\infty}^s T_0(s,\tau)P_0(\tau)g(\tau,v_\tau^{s,c})\,d\tau \\
&\quad - \int_s^{+\infty} \overline{T}(s,\tau)Q_0(\tau)g(\tau,v_\tau^{s,c})\,d\tau \\
&= P(s)v_s^{s,c} + Q(s)v_s^{s,c}.
\end{aligned} \tag{9.34}
$$

Note that

$$
v^{s,c} = v(\cdot, s, c + z(s,c)).
$$

We have $z(s,0) = 0$ since taking $\varphi \in C$ with $c = R(s)\varphi = 0$, the function $v = 0$ satisfies (9.13) (recall that $g(t,0) = 0$ for $t \in \mathbb{R}$) and so $z(s,0) = 0$. Moreover, by Lemma 9.1 we have

$$
\|v_t^{s,c} - v_t^{s,\bar{c}}\| \leq 2Ne^{(\mu+4\delta N)|t-s|}\|c - \bar{c}\|,
$$

which establishes (9.31). On the other hand, it follows from (9.34) that

$$\|z(s,c) - z(s,\bar{c})\| \le \delta N \int_{-\infty}^{s} e^{-\lambda(s-\tau)} \|v_{\tau}^{s,c} - \overline{v}_{\tau}^{s,\overline{c}}\| \, d\tau$$

$$+ \delta N \int_{s}^{+\infty} e^{\lambda(s-\tau)} \|v_{\tau}^{s,c} - \overline{v}_{\tau}^{s,\overline{c}}\| \, d\tau$$

$$\le 2\delta N^2 \int_{-\infty}^{s} e^{-\lambda(s-\tau)} e^{(\mu+4\delta N)(\tau-s)} \, d\tau \|c - \overline{c}\|$$

$$+ 2\delta N^2 \int_{s}^{+\infty} e^{\lambda(s-\tau)} e^{(\mu+4\delta N)(\tau-s)} \, d\tau \|c - \overline{c}\|$$

$$\le 2\delta N^2 \left(\frac{1}{\lambda + \mu + 4\delta N} + \frac{1}{\lambda - \mu - 4\delta N} \right) \|c - \overline{c}\|.$$

Hence, taking δ sufficiently small (recall that $\mu < \lambda$), we have $z \in \mathcal{C}^L$. Finally, one can show as in the proof of Theorem 7.3 that graph $z = V^c$. $\square$

9.4 Smooth Center Manifolds

In this section we show that there exist smooth center invariant manifolds for Eq. (7.1) when the linear equation $v' = L(t)v_t$ has an exponential trichotomy and g is a sufficiently small C^k perturbation. We assume that:

(1) $L(t) \colon C \to \mathbb{R}^n$, for $t \in \mathbb{R}$, are bounded linear operators such that the map $(t, v) \mapsto L(t)v$ is of class C^k on $\mathbb{R} \times C$ and condition (2.22) holds;
(2) $g \colon \mathbb{R} \times C \to \mathbb{R}^n$ is of class C^k, and $g(t, 0) = 0$ and $\partial g(t, 0) = 0$ for $t \in \mathbb{R}$.

Here and everywhere else in this section, ∂ shall denote the partial derivative with respect to the last variable of each function.

Assume that Eq. (2.12) has an exponential trichotomy. Let $\mathcal{C}^{k+L}$ be the set of all functions $z \in \mathcal{C}^L$ (see Sec. 9.3) of class C^k in c such that $\partial z(s, 0) = 0$, $\|\partial^j z(s, c)\| \le 1$ for $j = 1, \ldots, k$ and

$$\|\partial^k z(s, c) - \partial^k z(s, \bar{c})\| \le \|c - \overline{c}\| \tag{9.35}$$

for every $s \in \mathbb{R}$ and $c, \overline{c} \in G(s)$. Note that a function $z \in \mathcal{C}^L$ is in $\mathcal{C}^{k+L}$ if and only if it is of class C^k in c and for each $s \in \mathbb{R}$:

(1) $z(s, 0) = 0$, $\partial z(s, 0) = 0$ and $z(s, G(s)) \subset E(s) \oplus F(s)$;
(2) for $c, \overline{c} \in G(s)$ we have $\|\partial^j z(s, c)\| \le 1$ for $j = 1, \ldots, k$ and (9.35) holds.

The graph of a function $z \in \mathcal{C}^{k+L}$ is given by (9.30).

The following result is a smooth center manifold theorem for Eq. (7.1). It requires the spectral gap condition $\lambda > (k + 1)\mu$.

Theorem 9.3 (Smooth center manifold). *Assume that Eq. (2.12) has an exponential trichotomy and that there exists $\delta > 0$ such that (8.29) holds for every $t \in \mathbb{R}$, $u,v \in C$ and $j = 1,\dots,k$. If $\lambda > (k+1)\mu$ and δ is sufficiently small, then the function $z \in \mathcal{C}^L$ given by Theorem 9.2 is in $\mathcal{C}^{k+L}$. Moreover, provided that δ is sufficiently small, for every $t,s \in \mathbb{R}$, $c,\bar{c} \in G(s)$ and $j = 0,\dots,k$ we have*

$$\left\| \frac{\partial^j w_t^c}{\partial c^j} - \frac{\partial^j w_t^{\bar{c}}}{\partial c^j} \right\| \leq 2Ne^{j(\mu+4\delta N)(t-s)} \|c - \bar{c}\|,$$

where $w_t^c = v_t(\cdot, s, c + z(s,c))$.

Proof. We divide the proof into steps.

Step 1. Auxiliary spaces and fixed points

We consider the set $\mathcal{Z}_+$ of all continuous functions

$$v \colon \big\{ (t,s,c) : t \in [s-r,+\infty), s \in \mathbb{R}, c \in G(s) \big\} \to \mathbb{R}^n$$

such that

$$\|v_t^{s,c}\| \leq 2Ne^{(\mu+4\delta N)(t-s)} \|c\| \tag{9.36}$$

for $t \geq s$, where $v^{s,c} = v(\cdot, s, c)$. It is a complete metric space when equipped with the norm

$$\|v_t\|_{\mathcal{Z}_+} = \sup\left\{ \frac{\|v_t^{s,c}\|}{\|c\|} e^{-(\mu+4\delta N)(t-s)} : t \in [s,+\infty), s \in \mathbb{R}, c \in G(s) \setminus \{0\} \right\}. \tag{9.37}$$

Similarly, we consider the set $\mathcal{Z}_-$ of all continuous functions

$$v \colon \big\{ (t,s,c) : t \in (-\infty,s], s \in \mathbb{R}, c \in G(s) \big\} \to \mathbb{R}^n$$

such that

$$\|v_t^{s,c}\| \leq 2Ne^{(\mu+4\delta N)(s-t)} \|c\|$$

for $t \leq s$, where $v^{s,c} = v(\cdot, s, c)$. It is a complete metric space when equipped with the norm

$$\|v_t\|_{\mathcal{Z}_-} = \sup\left\{ \frac{\|v_t^{s,c}\|}{\|c\|} e^{-(\mu+4\delta N)(s-t)} : t \in (-\infty,s], s \in \mathbb{R}, c \in G(s)) \setminus \{0\} \right\}. \tag{9.38}$$

In a similar manner to that in the proof of Theorem 9.2, one can show that the operator $J\colon \mathcal{Z}_+ \to \mathcal{Z}_+$ defined by

$$
\begin{aligned}
(Jv)_t^{s,c} = T(t,s)c &+ \int_s^t T(t,\tau)R_0(\tau)g(\tau,v_\tau^{s,c})\,d\tau \\
&+ \int_{-\infty}^t T_0(t,\tau)P_0(\tau)g(\tau,v_\tau^{s,c})\,d\tau \\
&- \int_t^{+\infty} \overline{T}(t,\tau)Q_0(\tau)g(\tau,v_\tau^{s,c})\,d\tau
\end{aligned}
\tag{9.39}
$$

has a unique fixed point $v_+ \in \mathcal{Z}_+$. Similarly, there exists a unique function $v_- \in \mathcal{Z}_-$ satisfying (9.32) for all $s \in \mathbb{R}$, $c \in G(s)$ and $t \le s$. Moreover, given $(s,c) \in \mathbb{R} \times G(s)$, the functions $v_+^{s,c} \in \mathcal{Z}_+^c$ and $v_-^{s,c} \in \mathcal{Z}_-^c$ are, respectively, the unique solutions of (9.13) and (9.14) with $R(s)\varphi = c$.

We also consider the set $\mathcal{Z}_+^{k+L}$ (respectively $\mathcal{Z}_-^{k+L}$) of functions $v \in \mathcal{Z}_+$ (respectively $\mathcal{Z}_-$) of class C^k in c such that

$$
\|\partial^j v_t^{s,c}\|e^{-j(\mu+4\delta N)|t-s|} \le 2N \quad \text{for } j = 1,\dots,k
\tag{9.40}
$$

and

$$
\|\partial^k v_t^{s,c} - \partial^k v_t^{s,\bar{c}}\|e^{-(k+1)(\mu+4\delta N)|t-s|} \le 2N\|c-\bar{c}\|
\tag{9.41}
$$

for $t \ge s-r$ (respectively $t \le s$), $s \in \mathbb{R}$ and $c,\bar{c} \in G(s)$, where ∂ denotes the derivative with respect to c. By Lemma 8.3, $\mathcal{Z}_+^{k+L}$ and $\mathcal{Z}_-^{k+L}$ are complete metric spaces when equipped with the distances in (9.37) and (9.38).

In the remainder of the proof we consider only the case of $t \ge s$. The case of $t \le s$ can be treated similarly. Let $v_+ \in \mathcal{Z}_+$ be the fixed point of the operator $J\colon \mathcal{Z}_+ \to \mathcal{Z}_+$ in (9.39). We want to show that $v_+ \in \mathcal{Z}_+^{k+L}$. Since $\mathcal{Z}_+^{k+L}$ is a closed subset of $\mathcal{Z}_+$, it suffices to show that $J(\mathcal{Z}_+^{k+L}) \subset \mathcal{Z}_+^{k+L}$.

Step 2. Estimating the derivatives

Now we obtain bounds involving the derivatives of the function

$$
h(t,s,c) = g(t,v_t^{s,c}).
$$

We first recall the Faà di Bruno formula for the mth derivative of a composition (see [29]). Consider open sets Y, Z and W of Banach spaces and assume that

$$
f_1\colon Y \to Z \quad \text{and} \quad f_0\colon Z \to W
$$

have derivatives up to order m. The mth derivative of $h = f_0 \circ f_1$ at a point $x \in Y$ is given by

$$d_x^m h = \sum_{\ell=1}^{m} d_y^k f_0 \sum_{\substack{0 \le r_1,\dots,r_\ell \le m \\ r_1+\dots+r_\ell=m}} c_{r_1 \cdots r_\ell} d_x^{r_1} f_1 \cdots d_x^{r_\ell} f_1$$

for some integers $c_{r_1 \cdots r_\ell} \ge 0$. There exists $\nu_1 = \nu_1(m) > 0$ such that

$$\|d_x^m h\| \le \nu_1 \sum_{\ell=1}^{m} \|d_y^\ell f_0\| \sum_{p(m,\ell)} \prod_{j=1}^{m} \|d_x^j f_1\|^{\ell_j}, \qquad (9.42)$$

where

$$p(m,\ell) = \left\{ (\ell_1,\dots,\ell_m) \in \mathbb{N}_0^m : \sum_{j=1}^{m} \ell_j = \ell \text{ and } \sum_{j=1}^{m} j\ell_j = m \right\}. \qquad (9.43)$$

Moreover, for each $y = f_1(x)$ and $\bar{y} = f_1(\bar{x})$, we have

$$\|d_x^m h - d_{\bar{x}}^m h\| \le \nu_1 \sum_{\ell=1}^{m} \|d_y^\ell f_0 - d_{\bar{y}}^\ell f_0\| \sum_{p(m,\ell)} \prod_{j=1}^{m} \|d_x^j f_1\|^{\ell_j} + \nu_2 \sum_{\ell=1}^{m} \|d_{\bar{y}}^\ell f_0\| S_\ell$$

$$(9.44)$$

for some constant $\nu_2 = \nu_2(m) > 0$, where

$$S_\ell = \sum_{p(m,\ell)} \sum_{j=1}^{m} T_j \prod_{q=1}^{j-1} \|d_{\bar{x}}^q f_1\|^{\ell_q} \prod_{q=j+1}^{m} \|d_x^q f_1\|^{\ell_q}$$

and

$$T_j = \|d_x^j f_1 - d_{\bar{x}}^j f_1\| \sum_{\ell=0}^{\ell_j-1} \|d_x^j f_1\|^{\ell_j-1-\ell} \|d_{\bar{x}}^j f_1\|^{\ell}.$$

By (8.29), (9.42) and (9.43), there exists $\nu_3 = \nu_3(m) > 0$ such that

$$\|\partial^m h(t,s,c)\| \le \nu_1 \sum_{\ell=1}^{m} \|\partial^\ell g(t, v_t^{s,c})\| \sum_{p(m,\ell)} \prod_{j=1}^{m} \|\partial^j v_t^{s,c}\|^{\ell_j}$$

$$\le \nu_1 \delta \sum_{\ell=1}^{m} \sum_{p(m,\ell)} \prod_{j=1}^{m} (2N)^{\ell_j} e^{j\ell_j(\mu+4\delta N)|t-s|}$$

$$\le \nu_1 \delta \sum_{\ell=1}^{m} (2N)^\ell \sum_{p(m,\ell)} e^{\sum_{j=1}^{m} j\ell_j(\mu+4\delta N)|t-s|} \qquad (9.45)$$

$$\le \nu_3 \delta e^{m(\mu+4\delta N)|t-s|}$$

for $m = 1,\dots,k$ (again, ∂ is the derivative with respect to the last variable).

Moreover, by (9.44) together with (8.29) we obtain

$$\|\partial^m h(t,s,c) - \partial^m h(t,s,\bar{c})\|$$

$$\leq \nu_1 \delta \|v_t^{s,c} - v_t^{s,\bar{c}}\| \sum_{\ell=1}^{m} \sum_{p(m,\ell)} \prod_{j=1}^{m} \|\partial^j v_t^{s,c}\|^{\ell_j} + \nu_2 \sum_{\ell=1}^{m} \|\partial^\ell g(t, v_t^{s,\bar{c}})\| S_\ell$$

$$\leq \nu_1 \delta \|c - \bar{c}\| 2N e^{(\mu+4\delta N)|t-s|} \sum_{\ell=1}^{m} \sum_{p(m,\ell)} \prod_{j=1}^{m} (2N)^{\ell_j} e^{j\ell_j(\mu+4\delta N)|t-s|} \qquad (9.46)$$

$$+ \nu_2 \sum_{\ell=1}^{m} S_\ell.$$

We have

$$S_\ell = \sum_{p(m,\ell)} \sum_{j=1}^{m} T_j \prod_{q=1}^{j-1} \|\partial^q v_t^{s,\bar{c}}\|^{\ell_q} \prod_{q=j+1}^{m} \|\partial^q v_t^{s,c}\|^{\ell_q}$$

$$\leq \sum_{p(m,\ell)} \sum_{j=1}^{m} T_j \prod_{q=1}^{j-1} (2N)^{\ell_q} e^{q\ell_q(\mu+4\delta N)|t-s|} \prod_{q=j+1}^{m} (2N)^{\ell_q} e^{q\ell_q(\mu+4\delta N)|t-s|},$$

$$(9.47)$$

where

$$T_j \leq \|\partial^j v_t^{s,c} - \partial^j v_t^{s,\bar{c}}\| \sum_{\ell=0}^{\ell_j-1} \|\partial^j v_t^{s,c}\|^{\ell_j-1-\ell} \|\partial^j v_t^{s,\bar{c}}\|^{\ell}$$

$$\leq 2N e^{(j+1)(\mu+4\delta N)|t-s|} \|c - \bar{c}\| \sum_{\ell=0}^{\ell_j-1} (2N)^{\ell_j-1} e^{j(\ell_j-1)(\mu+4\delta N)|t-s|}$$

$$\leq \|c - \bar{c}\| \sum_{\ell=0}^{\ell_j-1} (2N)^{\ell_j} e^{(j\ell_j+1)(\mu+4\delta N)|t-s|}.$$

Hence, by (9.47) there exists $\nu_4 = \nu_4(m) > 0$ such that

$$S_\ell \leq \|c - \bar{c}\| \sum_{p(m,\ell)} \sum_{j=1}^{m} \sum_{\ell=0}^{\ell_j-1} (2N)^{\sum_{q=1}^{m} \ell_q} e^{(1+\sum_{q=1}^{m} q\ell_q)(\mu+4\delta N)|t-s|}$$

$$\leq \nu_4 \|c - \bar{c}\| (2N)^\ell e^{(m+1)(\mu+4\delta N)|t-s|}.$$

In a similar manner to that in (9.45), it follows from (9.46) that there exists a constant $\nu_5 = \nu_5(m) > 0$ such that

$$\|\partial^m h(t,s,c) - \partial^m h(t,s,\overline{c})\| \leq \nu_1 \delta \|c - \overline{c}\| e^{(m+1)(\mu+4\delta N)|t-s|} \sum_{\ell=1}^{m} (2N)^{\ell+1}$$

$$+ \nu_2 \nu_4 \delta \|c - \overline{c}\| e^{(m+1)(\mu+4\delta N)|t-s|} \sum_{\ell=1}^{m} (2N)^{\ell}$$

$$\leq \nu_5 \delta \|c - \overline{c}\| e^{(m+1)(\mu+4\delta N)|t-s|}.$$

$$(9.48)$$

Step 3. Smoothness of the fixed points

First we show that the three integrals in (9.39) are of class C^k in c. By (9.4), (9.5) and (9.45) we have

$$\|T_0(t,\tau)P_0(\tau)\partial^k h(\tau,s,c)\| \leq \nu_3 \delta N e^{-\lambda(t-\tau)} e^{k(\mu+4\delta N)(\tau-s)}$$

for $\tau \leq t$ and

$$\|\overline{T}(t,\tau)Q_0(\tau)\partial^k h(\tau,s,c)\| \leq \nu_3 \delta N e^{\lambda(t-\tau)} e^{k(\mu+4\delta N)(\tau-s)}$$

for $\tau \geq t$. Hence, for any sufficiently small δ such that $\lambda > k(\mu + 4\delta N)$, the integrals

$$\int_{-\infty}^{t} T_0(t,\tau)P_0(\tau)\partial^k h(\tau,s,c)\,d\tau, \quad \int_{t}^{+\infty} \overline{T}(s,\tau)Q_0(\tau)\partial^k h(\tau,s,c)\,d\tau$$

are well defined and converge uniformly for $c \in G(s)$. Since the derivative $\partial^k h(\tau,s,c)$ is continuous, it follows from the Leibniz rule for improper integrals that the second and third integrals in (9.39) are of class C^k in c, with

$$\frac{\partial^j}{\partial c^j} \int_{-\infty}^{t} T_0(t,\tau)P_0(\tau)g(\tau,v_\tau^{s,c})\,d\tau = \int_{-\infty}^{t} T_0(t,\tau)P_0(\tau)\partial^j h(\tau,s,c)\,d\tau$$

and

$$\frac{\partial^j}{\partial c^j} \int_{t}^{+\infty} \overline{T}(t,\tau)Q_0(\tau)g(\tau,v_\tau^{s,c})\,d\tau = -\int_{t}^{+\infty} \overline{T}(t,\tau)Q_0(\tau)\partial^j h(\tau,s,c)\,d\tau$$

for $j = 1, \ldots, k$. Similarly, the first integral in (9.39) satisfies

$$\frac{\partial^j}{\partial c^j} \int_{s}^{t} T(t,\tau)R_0(\tau)g(\tau,v_\tau^{s,c})\,d\tau = \int_{s}^{t} T(t,\tau)R_0(\tau)\partial^j h(\tau,s,c)\,d\tau$$

for $j = 1, \ldots, k$. Therefore,

$$\partial^j (Jv)_s^{s,c} = \chi T(t,s)R(s) + \int_s^t T(t,\tau)R_0(\tau)\partial^j h(\tau,s,c)\,d\tau$$

$$+ \int_{-\infty}^t T_0(t,\tau)P_0(\tau)\partial^j h(\tau,s,c)\,d\tau$$

$$- \int_t^{+\infty} \overline{T}(t,\tau)Q_0(\tau)\partial^j h(\tau,s,c)\,d\tau,$$

where

$$\chi = \begin{cases} 1 & \text{if } j = 1, \\ 0 & \text{otherwise.} \end{cases}$$

In view of the inequality $\lambda > (k+1)\mu$, we have

$$\|\partial^j (Jv)_t^{s,c}\| \leq \chi \|T(t,\tau)R(\tau)\|$$

$$+ \int_s^t \|T(t,\tau)R_0(\tau)\| \cdot \|\partial^j h(\tau,s,c)\|\,d\tau$$

$$+ \int_{-\infty}^t \|T_0(t,\tau)P_0(\tau)\| \cdot \|\partial^j h(\tau,s,c)\|\,d\tau$$

$$+ \int_t^{+\infty} \|\overline{T}(t,\tau)Q_0(\tau)\| \cdot \|\partial^j h(\tau,s,c)\|\,d\tau$$

$$\leq \chi N e^{\mu(t-s)} + \nu_4 \delta N \left(\int_s^t e^{\mu(t-\tau)} e^{j(\mu+4\delta N)(\tau-s)}\,d\tau \right.$$

$$+ \int_{-\infty}^t e^{-\lambda(t-\tau)} e^{j(\mu+4\delta N)(\tau-s)}\,d\tau$$

$$\left. + \int_t^{+\infty} e^{\lambda(t-\tau)} e^{j(\mu+4\delta N)(\tau-s)}\,d\tau \right)$$

$$\leq \chi N e^{\mu(t-s)} + \nu_3 \delta N e^{j(\mu+4\delta N)(t-s)}$$

$$\times \left(\int_s^t e^{(\mu - j(\mu+4\delta N))(t-\tau)}\,d\tau \right.$$

$$+ \int_{-\infty}^t e^{-(\lambda + j(\mu+4\delta N))(t-\tau)}\,d\tau$$

$$\left. + \int_t^{+\infty} e^{-(\lambda - j(\mu+4\delta N))(\tau-t)}\,d\tau \right)$$

$$\leq N e^{j(\mu+4\delta N)(t-s)}(\chi + \theta')$$

for $t \geq s$, where

$$\theta' = \nu_3 \delta \left(\frac{1}{j(\mu+4\delta N) - \mu} + \frac{1}{\lambda + j(\mu+4\delta N)} + \frac{1}{\lambda - j(\mu+4\delta N)} \right).$$

For $j > 1$ and δ sufficiently small we obtain

$$\|\partial^j (Jv)_t^{s,c}\| e^{-j(\mu+4\delta N)(t-s)} \le 2N.$$

Moreover, for $j = 1$, by (9.40) we have

$$\|\partial h(t,s,c)\| \le \|\partial g(t, v_t^{s,c})\| \cdot \|\partial v_t^{s,c}\|$$
$$\le 2\delta N e^{(\mu+\delta N)(t-s)}$$

for $t \ge s$ and so $\nu_3(1) = 2N$. Hence, for any sufficiently small δ we obtain

$$\|\partial (Jv)_t^{s,c}\| e^{-(\mu+4\delta N)(t-s)}$$
$$\le \frac{3}{2}N + 2N^2 \delta \left(\frac{1}{\lambda + \mu + 4\delta N} + \frac{1}{\lambda - \mu - 4\delta N} \right) \le 2N.$$

By (9.48) together with (9.4) and (9.5), for each $t \ge s$ and $c, \bar{c} \in G(s)$ we have

$$\|\partial^k (Jv)_t^{s,c} - \partial^k (Jv)_t^{s,\bar{c}}\|$$
$$\le \int_s^t \|T(t,\tau) R_0(\tau)\| \cdot \|\partial^k h(\tau, s, c) - \partial^k h(\tau, s, \bar{c})\| \, d\tau$$
$$+ \int_{-\infty}^t \|T_0(t,\tau) P_0(\tau)\| \cdot \|\partial^k h(\tau, s, c) - \partial^k h(\tau, s, \bar{c})\| \, d\tau$$
$$+ \int_t^{+\infty} \|\overline{T}(t,\tau) Q_0(\tau)\| \cdot \|\partial^k h(\tau, s, c) - \partial^k h(\tau, s, \bar{c})\| \, d\tau$$
$$\le \nu_5 \delta N \|c - \bar{c}\| \left(\int_s^t e^{\mu(t-\tau)} e^{(k+1)(\mu+4\delta N)(\tau-s)} \, d\tau \right.$$
$$+ \int_{-\infty}^t e^{-\lambda(t-\tau)} e^{(k+1)(\mu+4\delta N)(\tau-s)} \, d\tau$$
$$\left. + \int_t^{+\infty} e^{\lambda(t-\tau)} e^{(k+1)(\mu+4\delta N)(\tau-s)} \, d\tau \right)$$
$$\le \nu_5 \delta N \|c - \bar{c}\| e^{(k+1)(\mu+4\delta N)(t-s)}$$
$$\times \left(\int_s^t e^{(\mu-(k+1)(\mu+4\delta N))(t-\tau)} \, d\tau \right.$$
$$+ \int_{-\infty}^t e^{-(\lambda+(k+1)(\mu+4\delta N))(t-\tau)} \, d\tau$$
$$\left. + \int_t^{+\infty} e^{-(\lambda-(k+1)(\mu+4\delta N))(\tau-t)} \, d\tau \right)$$
$$= \theta'' N \|c - \bar{c}\| e^{(k+1)(\mu+4\delta N)(t-s)},$$

where

$$\theta'' = \frac{\nu_5 \delta}{(k+1)(\mu + 4\delta N) - \mu} + \frac{\nu_5 \delta}{\lambda + (k+1)(\mu + 4\delta N)}$$
$$+ \frac{\nu_5 \delta}{\lambda - (k+1)(\mu + 4\delta N)}.$$

Finally, taking δ sufficiently small such that $\theta'' \leq 2$ we obtain

$$\|\partial^k (Jv)_t^{s,c} - \partial^k (Jv)_t^{s,\overline{c}}\| e^{-(k+1)(\mu + 4\delta N)(t-s)} \leq 2N \|c - \overline{c}\|$$

for $t \geq s$. Therefore, $J(\mathcal{Z}_+^{k+L}) \subset \mathcal{Z}_+^{k+L}$. Since $\mathcal{Z}_+^{k+L}$ is closed in $\mathcal{Z}_+$, the fixed point v_+ of the operator J in (9.39) is in $\mathcal{Z}_+^{k+L}$. Similarly, the unique function $v_- \in \mathcal{Z}_-$ satisfying (9.32) is in $\mathcal{Z}_-^{k+L}$. This shows that the continuous function

$$v \colon \mathbb{R} \times \{(s,c) \in \mathbb{R} \times C : c \in G(s)\} \to \mathbb{R}^n$$

defined by

$$v(t,s,c) = \begin{cases} v_+(t,s,c) & \text{if } t \geq s, \\ v_-(t,s,c) & \text{if } t < s \end{cases}$$

or, equivalently, by $v(t,s,c) = v^{s,c}(t)$ with $v^{s,c}$ in (9.33), is of class C^k in c and satisfies (9.36), (9.40) and (9.41) for $t \in \mathbb{R}$.

Step 4. Smoothness of the graphs

Finally, we show that $z \in \mathcal{C}^{k+L}$. By (9.34) and Step 3, z is of class C^k in c. Moreover, using the bounds obtained in the former step we obtain

$$\|\partial^j z(s,c)\| \leq \int_{-\infty}^{s} \|T_0(s,\tau) P_0(\tau)\| \cdot \|\partial^j h(\tau, s, c)\| \, d\tau$$
$$+ \int_{s}^{+\infty} \|\overline{T}(s,\tau) Q_0(\tau)\| \cdot \|\partial^j h(\tau, s, c)\| \, d\tau$$
$$\leq \nu_5 \delta N \left(\int_{-\infty}^{s} e^{-\lambda(s-\tau)} e^{j(\mu + 4\delta N)(\tau - s)} \, d\tau \right.$$
$$\left. + \int_{s}^{+\infty} e^{\lambda(s-\tau)} e^{j(\mu + 4\delta N)(\tau - s)} \, d\tau \right)$$
$$\leq \nu_5 \delta N \left(\frac{1}{\lambda + j(\mu + 4\delta N)} + \frac{1}{\lambda - j(\mu + 4\delta N)} \right) \leq 1$$

for any sufficiently small δ. Similarly, we have

$$\|\partial_c^k z(s,a) - \partial_c^k z(s,\bar{a})\|$$

$$\leq \int_{-\infty}^{s} \|T_0(s,\tau)P_0(\tau)\| \cdot \|\partial^k h(\tau,s,c) - \partial^k h(\tau,s,c)\| \, d\tau$$

$$+ \int_{s}^{+\infty} \|\overline{T}(s,\tau)Q_0(\tau)\| \cdot \|\partial^k h(\tau,s,c) - \partial^k h(\tau,s,c)\| \, d\tau$$

$$\leq \nu_5 \delta N \|c - \bar{c}\| \left(\int_{-\infty}^{s} e^{-\lambda(s-\tau)} e^{(k+1)(\mu+4\delta N)(\tau-s)} \, d\tau \right.$$

$$+ \left. \int_{s}^{+\infty} e^{\lambda(s-\tau)} e^{(k+1)(\mu+4\delta N)(\tau-s)} \, d\tau \right)$$

$$\leq \nu_5 \delta N \|c - \bar{c}\| \left(\frac{1}{\lambda + (k+1)(\mu+4\delta N)} + \frac{1}{\lambda - (k+1)(\mu+4\delta N)} \right)$$

$$\leq \|c - \bar{c}\|$$

for any sufficiently small δ and so $z \in \mathcal{C}^{k+L}$. This completes the proof of the theorem. $\qquad\square$

Chapter 10

Spectral Theory

This chapter is dedicated to the study of a version of the Sacker–Sell spectrum for a nonautonomous linear delay equation. We start with a discussion of a few basic properties of an exponential dichotomy. In particular, we show that for an exponential dichotomy on the line the stable and unstable spaces are uniquely determined. The central result of the chapter is a description of all possible forms of the spectrum. We also show that to each connected component of the spectrum one can associate naturally a subspace whose Lyapunov exponents are contained in the connected component. Finally, we consider nonlinear perturbations of a linear delay equation and we give a condition under which the Lyapunov exponents of the nonlinear dynamics also belong to a connected component of the spectrum.

10.1 Structure of the Spectrum

In this section we study the notion of spectrum for Eq. (2.12), that is,

$$v' = L(t)v_t,$$

where $L(t) \colon C \to \mathbb{R}^n$, for $t \in \mathbb{R}$, are bounded linear operators satisfying the standing assumptions of Sec. 2.2. We shall always assume in this chapter that $r > 0$. Let $T(t,s)$ be the evolution family associated with Eq. (2.12).

Proposition 10.1. *Assume that Eq. (2.12) has an exponential dichotomy on $\mathbb{R}$. Then for each $s \in \mathbb{R}$ we have*

$$E(s) = \left\{ \varphi \in C : \sup_{t \geq s} \|T(t,s)\varphi\| < +\infty \right\}$$

and $F(s)$ consists of all $\varphi \in C$ for which there exists a function $v \colon (-\infty, s] \to \mathbb{R}^n$ with $v_s = \varphi$ such that $v_t = T(t,\tau)v_\tau$ for $s \geq t \geq \tau$ and $\sup_{t \leq s} |v(t)| < +\infty$.

Proof. By (3.23), for $v \in E(s)$ we have

$$\sup_{t \geq s} \|T(t,s)\varphi\| < +\infty. \tag{10.1}$$

Now assume that (10.1) holds. It follows from (3.23) that

$$\sup_{t \geq s} \|T(t,s)Q(s)\varphi\| < +\infty \tag{10.2}$$

since otherwise

$$\sup_{t \geq s} \|T(t,s)\varphi\| \geq \sup_{t \geq s} \|T(t,s)Q(s)\varphi\| - \sup_{t \geq s} \|T(t,s)P(s)\varphi\|$$

$$\geq \sup_{t \geq s} \|T(t,s)Q(s)\varphi\| - N\|\varphi\| = +\infty.$$

Moreover, for $t \geq s$ we have

$$\|Q(s)\varphi\| = \|\overline{T}(s,t)T(t,s)Q(s)\varphi\|$$

$$\leq Ne^{-\lambda(t-s)}\|T(t,s)Q(s)\varphi\|.$$

Hence, if $Q(s)\varphi \neq 0$, then

$$\sup_{t \geq s} \|T(t,s)Q(s)\varphi\| = +\infty,$$

which contradicts (10.2). Therefore, $Q(s)\varphi = 0$ and $\varphi \in E(s)$.

Now take $\varphi \in F(s)$ and consider the function $v \colon (-\infty, s] \to \mathbb{R}^n$ defined by $v_t = \overline{T}(t,s)\varphi$ for $t \leq s$. Then $v_t = T(t,\tau)v_\tau$ for $s \geq t \geq \tau$ and it follows from (3.23) that $\sup_{t \leq s}|v(t)| < +\infty$. Finally, take $\varphi \in C$ and $v \colon (-\infty, s] \to \mathbb{R}^n$ as in the statement of the proposition. Moreover, write $\varphi = x + y$ with $x \in E(s)$ and $y \in F(s)$. Then

$$x = P(s)\varphi = T(s,t)P(t)v_t$$

for $t \leq s$ and it follows from

$$\|x\| = \|T(s,t)P(t)v_t\| \leq Ne^{-\lambda(s-t)}\|v_t\|$$

that $\sup_{t \leq s}|v(t)| = +\infty$ when $x \neq 0$. Hence, $x = 0$ and so $\varphi \in F(s)$. $\square$

A simple consequence of Proposition 10.1 is that for exponential dichotomies on $\mathbb{R}$ the stable and unstable spaces are uniquely determined. We note that in general this need not be the case for exponential dichotomies on an interval $I \neq \mathbb{R}$.

The spectrum of a linear delay equation is defined in terms of the notion of an exponential dichotomy for an arbitrary evolution family (see Definition 9.2).

Definition 10.1. The *spectrum* of Eq. (2.12) is the set Σ of all numbers $a \in \mathbb{R}$ such that the evolution family

$$T_a(t,s) = e^{-a(t-s)}T(t,s)$$

does not have an exponential dichotomy on $\mathbb{R}$.

This means that if $T(t,s)\varphi = v_t$, where v is the unique solution of Eq. (2.12) on the interval $[s - r, +\infty)$ with $v_s = \varphi$, then $T_a(t,s)\varphi = e^{-a(t-s)}v_t$, that is,

$$(T_a(t,s)\varphi)(\theta) = e^{-a(t-s)}v(t+\theta) \quad \text{for } \theta \in [-r, 0].$$

We emphasize that the evolution family $T_a(t,s)$ may not come from a linear delay equation. Indeed, if $w_t = T_a(t,s)\varphi$ for the solution w of some delay equation with $w_s = \varphi$, then since $w_t(\theta) = w_{t+\theta}(0)$, we should have

$$\begin{aligned}
e^{-a(t-s)}v(t+\theta) = (T_a(t,s)\varphi)(\theta) &= w_t(\theta) \\
&= w_{t+\theta}(0) = (T_a(t+\theta, s)\varphi)(0) \\
&= e^{-a(t+\theta-s)}v(t+\theta).
\end{aligned}$$

But in general this fails for $\theta \neq 0$. So in particular one cannot apply Proposition 10.1 to deduce that the stable and unstable spaces are uniquely determined, neither we know a priori whether the unstable spaces are finite-dimensional.

Proposition 10.2. *For each $a \in \Sigma \setminus \mathbb{R}$ the following properties hold:*

(1) the stable and unstable spaces $E_a(s)$ and $F_a(s)$ associated with the exponential dichotomy of the evolution family $T_a(t,s)$ are uniquely determined;

(2) $\dim F_a(s) < +\infty$ for all $s \in \mathbb{R}$.

Proof. Given $a \in \mathbb{R}$, we define an evolution family $\hat{T}_a(t,s) \colon C \to C$ as follows. For each solution v of Eq. (2.12) on the interval $[s - r, +\infty)$ we consider the function $w(t) = e^{-at}v(t)$ on the same interval. Then $\hat{T}_a(t,s)$ is the unique linear operator such that $\hat{T}_a(t,s)w_s = w_t$ for $t \geq s$.

We show that $\hat{T}_a(t,s)$ and $T_a(t,s)$ are conjugate by a linear map independent of t and s. Consider the map $H \colon C \to C$ defined by $H(\varphi) = \psi$, where $\psi(\theta) = e^{a\theta}\varphi(\theta)$. Then H is invertible and it follows readily from the definitions that

$$\hat{T}_a(t,s) = H^{-1} \circ T_a(t,s) \circ H$$

for each $t \geq s$. Moreover, since

$$e^{-|ar|}\|\varphi\| \leq \|\psi\| \leq e^{|ar|}\|\varphi\|,$$

we have

$$\|H\| \leq e^{|ar|} \quad \text{and} \quad \|H^{-1}\| \leq e^{|ar|}.$$

This implies that $\hat{T}_a(t,s)$ has an exponential dichotomy on $\mathbb{R}$ if and only if the same happens to $T_a(t,s)$, with the same constant λ. Moreover, the stable and unstable spaces $\hat{E}_a(s)$ and $\hat{F}_a(s)$ associated with the exponential dichotomy of $\hat{T}_a(t,s)$ satisfy

$$\hat{E}_a(s) = H^{-1}E_a(s) \quad \text{and} \quad \hat{F}_a(s) = H^{-1}F_a(s). \tag{10.3}$$

Now we show that the evolution family $\hat{T}_a(t,s)$ is obtained from a linear delay equation. First note that

$$w'(t) = -ae^{-at}v(t) + e^{-at}v'(t) = -aw(t) + e^{-at}L(t)v_t. \tag{10.4}$$

Writing $L(t)$ as the Riemann–Stieltjes integral in (2.23), we have

$$e^{-at}L(t)v_t = \int_{-r}^{0} d\eta(t,\theta)e^{-at}v(t+\theta) = \int_{-r}^{0} d\eta(t,\theta)e^{a\theta}w(t+\theta).$$

Therefore, it follows from (10.4) that

$$w' = L_a(t)w_t, \tag{10.5}$$

where

$$L_a(t)\varphi = \int_{-r}^{0} d\eta(t,\theta)e^{a\theta}\varphi(\theta) - a\varphi(0).$$

By (2.24), letting $\psi(\theta) = e^{a\theta}\varphi(\theta)$ we obtain

$$\|L_a(t)\varphi\| \leq \|L(t)\| \cdot \|\psi\| + \|a\varphi(0)\|$$
$$\leq \|L(t)\|e^{|ar|}\|\varphi\| + |a| \cdot \|\varphi\|$$

and so

$$\|L_a(t)\| \leq \|L(t)\|e^{|ar|} + |a|.$$

Since the evolution family $\hat{T}_a(t,s)$ is obtained from Eq. (10.5), it follows from Proposition 10.1 that for each $s \in \mathbb{R}$ we have

$$\hat{E}_a(s) = \left\{ \varphi \in C : \sup_{t \geq s}\|\hat{T}_a(t,s)\varphi\| < +\infty \right\}$$

and $\hat{F}_a(s)$ is the set of all $\varphi \in C$ for which there exists a function $w : (-\infty, s] \to \mathbb{R}^n$ with $w_s = \varphi$ such that $w_t = \hat{T}_a(t,\tau)w_\tau$ for $s \geq t \geq \tau$ and $\sup_{t \leq s}|w(t)| < +\infty$. This shows that the spaces $\hat{E}_a(s)$ and $\hat{F}_a(s)$ are uniquely determined and so, by (10.3), the same happens to $E_a(s)$ and $F_a(s)$. This establishes the first property.

For the second property, we note that by Theorem 3.2 the spaces $\hat{F}_a(s)$ are finite-dimensional. Hence, it follows from (10.3) and the invertibility along the unstable spaces in the notion of an exponential dichotomy that $F_a(t)$ is finite-dimensional for all $t \in \mathbb{R}$ (with a dimension that is independent of t). □

The following result describes all possible forms of the spectrum. Given $-\infty \le a \le b \le +\infty$, we write $|a, b| = \mathbb{R} \cap [a, b]$, that is,

$$|a, b| = \begin{cases} [a, b] & \text{if } a, b \in \mathbb{R}, \\ (-\infty, b] & \text{if } a = -\infty, b \in \mathbb{R}, \\ [a, +\infty) & \text{if } a \in \mathbb{R}, b = +\infty, \\ \mathbb{R} & \text{if } a = -\infty, b = +\infty. \end{cases}$$

Theorem 10.1 (Spectrum). *Assume that* $r > 0$. *For the spectrum of Eq. (2.12), one of the following alternatives holds:*

(1) $\Sigma = \emptyset$;
(2) $\Sigma = \bigcup_{m=1}^{k} |a_m, b_m|$ *for some numbers*

$$+\infty \ge b_1 \ge a_1 > b_2 \ge a_2 > \cdots > b_k \ge a_k \ge -\infty$$

with $k \in \mathbb{N}$ *(when* $k = 1$, $a_1 = -\infty$ *and* $b_1 = +\infty$ *we have* $\Sigma = \mathbb{R}$);
(3) $\Sigma = \bigcup_{m=1}^{\infty} |a_m, b_m|$ *for some numbers*

$$+\infty \ge b_1 \ge a_1 > b_2 \ge a_2 > b_3 \ge a_3 > \cdots \tag{10.6}$$

with $a_m \to -\infty$ *when* $m \to \infty$;
(4) $\Sigma = \bigcup_{m=1}^{\infty} |a_m, b_m| \cup (-\infty, a_\infty]$ *for some numbers as in* (10.6) *with* $a_m \to a_\infty \in \mathbb{R}$ *when* $m \to \infty$.

Proof. We first establish some auxiliary results.

Lemma 10.1. *The set* $\Sigma \subset \mathbb{R}$ *is closed and for each* $a \in \mathbb{R} \setminus \Sigma$ *we have*

$$E_a(t) = E_b(t) \quad \text{and} \quad F_a(t) = F_b(t) \quad \text{for all } t \in \mathbb{R} \tag{10.7}$$

and all b *in some open neighborhood of* a.

Proof of the lemma. Given $a \in \mathbb{R} \setminus \Sigma$, the following properties hold:

(1) there exist projections $P(t) \colon C \to C$ for $t \in \mathbb{R}$ satisfying (3.2);
(2) the linear operator

$$T_a(t, s)| \operatorname{Ker} P(s) \colon \operatorname{Ker} P(s) \to \operatorname{Ker} P(t)$$

is invertible for $t \ge s$ and so the same happens to the linear operator in (3.3);
(3) there exist constants $\lambda, D > 0$ such that

$$\|e^{-a(t-s)} T(t, s) P(s)\| \le D e^{-\lambda(t-s)}$$

and

$$\|e^{-a(s-t)} \overline{T}(s, t) Q(t)\| \le D e^{-\lambda(t-s)}$$

for $t, s \in \mathbb{R}$ with $t \ge s$, where $Q(t) = \operatorname{Id} - P(t)$.

Therefore, for each $b \in \mathbb{R}$ we obtain

$$\|e^{-b(t-s)}T(t,s)P(s)\| \leq De^{-(\lambda-a+b)(t-s)}$$

and

$$\|e^{-b(s-t)}\overline{T}(s,t)Q(t)\| \leq De^{-(\lambda+a-b)(t-s)}$$

for $t \geq s$. Hence, $b \in \mathbb{R}\backslash\Sigma$ whenever $|a-b| < \lambda$ and so Σ is closed. Moreover, by Proposition 10.1, the stable and unstable spaces are uniquely determined and since the former four inequalities use the same pair of projections, we conclude that (10.7) holds whenever $|a - b| < \lambda$. $\qquad\square$

Lemma 10.2. *Take $a, b \in \mathbb{R} \setminus \Sigma$ with $a < b$. Then $[a,b] \cap \Sigma \neq \emptyset$ if and only if $\dim F_a(t) > \dim F_b(t)$.*

Proof of the lemma. Note that

$$E_a(s) \subset E_b(s) \quad \text{and} \quad F_b(s) \subset F_a(s), \qquad\qquad (10.8)$$

as well as

$$C = E_a(s) \oplus F_a(s) \qquad\qquad (10.9)$$

for $s \in \mathbb{R}$. Now assume that

$$\dim F_a(t) = \dim F_b(t)$$

(since $a, b \in \mathbb{R} \setminus \Sigma$, the two dimensions are finite). Then (10.7) holds, in view of (10.8) and (10.9). Hence, there exist projections $P(t)$ for $t \in \mathbb{R}$ satisfying (3.2) and there exist constants $\lambda, \mu, D > 0$ such that

$$\|e^{-a(t-s)}T(t,s)P(s)\| \leq De^{-\lambda(t-s)}, \quad \|e^{-b(t-s)}T(t,s)P(s)\| \leq De^{-\mu(t-s)}$$
$$(10.10)$$

and

$$\|e^{-a(s-t)}\overline{T}(s,t)Q(t)\| \leq De^{-\lambda(t-s)}, \quad \|e^{-b(s-t)}\overline{T}(s,t)Q(t)\| \leq De^{-\mu(t-s)}$$
$$(10.11)$$

for $t, s \in \mathbb{R}$ with $t \geq s$ (with the same observation as in the proof of Lemma 10.1 on the invertibility of the linear operator in (3.3)). For each $c \in [a, b]$, by (10.10) and (10.11) we obtain

$$\|e^{-c(t-s)}T(t,s)P(s)\| \leq De^{-\lambda(t-s)} \leq De^{-\min\{\lambda,\mu\}(t-s)}$$

and

$$\|e^{-c(s-t)}\overline{T}(s,t)Q(t)\| \leq De^{-\mu(t-s)} \leq De^{-\min\{\lambda,\mu\}(t-s)}$$

for $t \geq s$. Therefore, the evolution family $T_c(t,s)$ has an exponential dichotomy on $\mathbb{R}$. Hence, $[a,b] \subset \mathbb{R} \setminus \Sigma$.

Now assume that $\dim F_a(t) > \dim F_b(t)$ and let

$$d = \inf\{c \in \mathbb{R} \setminus \Sigma : \dim F_c(t) = \dim F_b(t)\}.$$

Since $\dim F_a(t) > \dim F_b(t)$, it follows from Lemma 10.1 that $a < d < b$. We want to show that $d \in \Sigma$. Assuming otherwise, we consider the cases

$$\dim F_d(t) = \dim F_b(t) \quad \text{and} \quad \dim F_d(t) \neq \dim F_b(t).$$

In the first case, by (10.8) and Lemma 10.1, there exists $\varepsilon > 0$ such that $\dim F_c(t) = \dim F_b(t)$ and $c \in \mathbb{R} \setminus \Sigma$ for $c \in (d - \varepsilon, d]$. But this contradicts the definition of d. In the second case, again by (10.8) and Lemma 10.1, there exists $\varepsilon > 0$ such that

$$\dim F_c(t) \neq \dim F_b(t)$$

and $c \in \mathbb{R} \setminus \Sigma$ for $c \in [d, d + \varepsilon)$, which again contradicts the definition of d. Hence, $d \in \Sigma$ and $[a, b] \cap \Sigma \neq \emptyset$. $\square$

Lemma 10.3. *For each $c \in \mathbb{R} \setminus \Sigma$, the set $\Sigma \cap [c, +\infty)$ is the union of finitely many disjoint closed intervals or is the empty set.*

Proof of the lemma. Let $d = \dim F_c(t)$ and assume that the intersection $\Sigma \cap [c, +\infty)$ has at least $d + 2$ connected components, say

$$I_i = |\alpha_i, \beta_i|, \quad \text{for } i = 1, \ldots, d+2,$$

where

$$\alpha_1 \leq \beta_1 < \alpha_2 \leq \beta_2 < \cdots < \alpha_{d+2} \leq \beta_{d+2} \leq +\infty.$$

For $i = 1, \ldots, d+1$, take $c_i \in (\beta_i, \alpha_{i+1})$. It follows from Lemma 10.2 that

$$d > \dim F_{c_1}(t) > \dim F_{c_2}(t) > \cdots > \dim F_{c_{d+1}}(t),$$

which is impossible since all the dimensions are nonnegative integers. $\square$

We proceed with the proof of the theorem. Since Σ is closed, if it is nonempty and has finitely many connected components, then it satisfies alternative 2.

Now we assume that Σ has infinitely many connected components. Take $c_1 \in \mathbb{R} \setminus \Sigma$ such that $\Sigma \cap [c_1, +\infty)$ is nonempty. By Lemma 10.3, $\Sigma \cap [c_1, +\infty)$ is the union of finitely many disjoint closed intervals, say $I_1, \ldots, I_k$. Note that $\Sigma \cap (-\infty, c_1) \neq \emptyset$ since otherwise

$$\Sigma = I_1 \cup \cdots \cup I_k,$$

which contradicts our assumption. Moreover, there exists $c_2 < c_1$ such that $c_2 \notin \Sigma$ and $(c_2, c_1) \cap \Sigma \neq \emptyset$. Otherwise,

$$\Sigma \cap (-\infty, c_1) = (-\infty, a]$$

for some $a < c_1$ and thus,

$$\Sigma = (-\infty, a] \cup I_1 \cdots \cup I_k,$$

which again contradicts our assumption. Proceeding inductively, we obtain a decreasing sequence $(c_m)_{m \in \mathbb{N}}$ of real numbers such that

$$c_m \notin \Sigma \quad \text{and} \quad \Sigma \cap (c_{m+1}, c_m) \neq \emptyset$$

for all $m \in \mathbb{N}$. Now we consider the cases

$$\lim_{m \to +\infty} c_m = -\infty \quad \text{and} \quad \lim_{m \to +\infty} c_m = a_\infty,$$

where $a_\infty \in \mathbb{R}$. In the first case, by Lemma 10.3 the spectrum Σ is given by alternative 3. In the second case, it follows from Lemma 10.3 that

$$\Sigma \cap (a_\infty, +\infty) = \bigcup_{m=1}^{\infty} |a_m, b_m|$$

for some numbers a_m and b_m as in (10.6) with $a_m \to a_\infty$ when $m \to \infty$. Again by Lemma 10.3, we have $(-\infty, a_\infty] \subset \Sigma$ and so Σ is given by the last alternative in the theorem. $\square$

We also consider briefly the particular case of linear equations with bounded growth (see (2.21)).

Proposition 10.3. *If Eq. (2.12) has bounded growth, then its spectrum satisfies $\Sigma \subset (-\infty, \omega]$.*

Proof. We have

$$\|T(t, s)\| \leq K e^{\omega(t-s)} \quad \text{for } t \geq s.$$

Hence, given $a > \omega$, we obtain

$$\|T_a(t, s)\| \leq K e^{(\omega - a)(t-s)} \quad \text{for } t \geq s$$

and so Eq. (2.12) has an exponential contraction on $\mathbb{R}$. This yields the desired result. $\square$

10.2   Lyapunov Exponents

In this section we assume that the spectrum Σ is nonempty and we associate spaces $G_I(s)$, for $s \in \mathbb{R}$, to each connected component $I \subset \Sigma$. It turns out that the Lyapunov exponents of the initial conditions in $G_I(s)$ belong to I. Let

$$E_{+\infty}(s) = F_{-\infty}(s) = C$$

for each $s \in \mathbb{R}$. We consider two cases:

(1) When Σ is given by alternative 2 in Theorem 10.1, for each connected component $I = \,]a_m, b_m[$, with $m = 1, \ldots, k$, we define

$$G_I(s) = E_{c_{m-1}}(s) \cap F_{c_m}(s), \tag{10.12}$$

where $c_m \in (b_{m+1}, a_m)$ for $m = 1, \ldots, k-1$,

$$c_0 \geq b_1 + \delta \quad \text{and} \quad c_k \leq b_m - \delta$$

with $\delta > 0$. Note that $c_0 = +\infty$ when $b_1 = +\infty$ and that $c_k = -\infty$ when $b_m = -\infty$.

(2) When Σ is given by alternatives 3 or 4 in Theorem 10.1, for each connected component $I = \,]a_m, b_m[$, with $m \in \mathbb{N}$, we define $G_I(s)$ as in (10.12), where

$$c_m \in (b_{m+1}, a_m) \text{ for } m \in \mathbb{N} \quad \text{and} \quad c_0 \geq b_1 + \delta$$

with $\delta > 0$. Finally, for the connected component $I = (-\infty, a_\infty]$ in alternative 4, we define

$$G_I(s) = \bigcap_{m \in \mathbb{N}} E_{c_m}(s).$$

It follows from Lemma 10.2 that the spaces $G_I(s)$ are independent of the choice of numbers c_m and δ.

We first describe decompositions of C in terms of the spaces $G_I(s)$.

Proposition 10.4. *Assume that $r > 0$ and that Σ is nonempty. Then*

$$C = E_{c_m}(s) \oplus \bigoplus_{p=1}^{m} G_{]a_p, b_p[}(s) \tag{10.13}$$

for each m and $s \in \mathbb{R}$. Moreover, if Σ has finitely many connected components, then

$$C = \bigoplus_{I \subset \Sigma} G_I(s) \quad \text{for } s \in \mathbb{R}. \tag{10.14}$$

Proof. Since

$$(A_1 + A_2) \cap A_3 = A_1 + (A_2 \cap A_3)$$

whenever A_i, for $i = 1, 2, 3$, are subspaces of C with $A_1 \subset A_3$, we have

$$
\begin{aligned}
C &= E_{c_1}(s) \oplus F_{c_1}(s) \\
&= \big((E_{c_2}(s) \oplus F_{c_2}(s)) \cap E_{c_1}(s)\big) \oplus F_{c_1}(s) \\
&= E_{c_2}(s) \oplus \big(F_{c_2}(s) \cap E_{c_1}(s)\big) \oplus F_{c_1}(s)
\end{aligned}
$$

and identity (10.13) can be obtained in finitely many steps, using the fact that $F_{c_1}(s) = G_{|a_1,b_1|}(s)$ since $E_{c_0}(s) = C$. On the other hand, (10.14) follows from $G_{|a_k,b_k|}(s) = E_{c_{k-1}}(s)$, because $F_{c_k}(s) = C$. $\qquad\square$

Now we show that the Lyapunov exponents of the initial conditions in the space $G_I(s)$ belong to I.

Theorem 10.2. *Assume that $r > 0$ and that Σ is nonempty. For each $s \in \mathbb{R}$ and $\varphi \in G_I(s) \setminus \{0\}$, we have*

$$\left(\liminf_{t \to +\infty} \frac{1}{t} \log \|T(t,s)\varphi\|, \limsup_{t \to +\infty} \frac{1}{t} \log \|T(t,s)\varphi\| \right) \subset I$$

and there exists a function $v \colon (-\infty, s] \to \mathbb{R}^n$ with $v_s = \varphi$ such that

$$v_\tau = T(\tau, \sigma) v_\sigma \quad \text{for } s \geq \tau \geq \sigma$$

and

$$\left(\liminf_{t \to -\infty} \frac{1}{t} \log \|v_t\|, \limsup_{t \to -\infty} \frac{1}{t} \log \|v_t\| \right) \subset I.$$

Proof. We consider only the case when I is bounded. The general case is obtained considering only one of the lower and upper bounds. Since $c_{m-1} \notin \Sigma$, the evolution family $T_{c_{m-1}}(t,s)$ has an exponential dichotomy and so there exist projections $P(t)$ for $t \in \mathbb{R}$ satisfying (3.2) and there exist constants $\lambda, D > 0$ such that

$$\|T(t,s)P(s)\| \leq De^{(c_{m-1}-\lambda)(t-s)} \tag{10.15}$$

and

$$\|\overline{T}(s,t)Q(t)\| \leq De^{-(\lambda+c_{m-1})(t-s)}$$

for $t \geq s$, where $Q(t) = \mathrm{Id} - P(t)$. By Proposition 10.1, we have $\mathrm{Im}\,P(t) = E_{c_{m-1}}(t)$ for $t \in \mathbb{R}$. Hence, each $\varphi \in G_{|a_m,b_m|}$ belongs to $\mathrm{Im}\,P(s)$ and so, by (10.15),

$$\limsup_{t \to +\infty} \frac{1}{t} \log \|T(t,s)\varphi\| \leq c_{m-1} - \lambda < c_{m-1}.$$

Letting $c_{m-1} \searrow b_m$, we obtain

$$\limsup_{t \to +\infty} \frac{1}{t} \log\|T(t,s)\varphi\| \le b_m.$$

Similarly, since $c_m \notin \Sigma$, there exist projections $\overline{P}(t)$ for $t \in \mathbb{R}$ satisfying (3.2) and there exist constants $\mu, D > 0$ such that

$$\|T(t,s)\overline{P}(s)\| \le De^{(c_m-\mu)(t-s)}$$

and

$$\|\overline{T}(s,t)\overline{Q}(t)\| \le De^{-(\mu+c_m)(t-s)} \tag{10.16}$$

for $t \ge s$, where $\overline{Q}(t) = \mathrm{Id} - \overline{P}(t)$. By Proposition 10.1, we have $\mathrm{Im}\,\overline{Q}(t) = F_{c_m}(t)$ for $t \in \mathbb{R}$. Hence, each $\varphi \in G_{|a_m,b_m|}$ belongs to $\mathrm{Im}\,\overline{Q}(s)$ and so, by (10.16),

$$\|v\| \le De^{-(\mu+c_m)(t-s)}\|\overline{T}(t,s)\varphi\| \quad \text{for } t \ge s.$$

Hence,

$$\liminf_{t \to +\infty} \frac{1}{t} \log\|T(t,s)\varphi\| \ge \mu + c_m > c_m$$

and letting $c_m \nearrow a_m$ yields the inequality

$$\liminf_{t \to +\infty} \frac{1}{t} \log\|T(t,s)\varphi\| \ge a_m.$$

This completes the proof of the first statement in the theorem. A similar argument yields the corresponding statement for negative time. $\qquad\square$

10.3 One-Sided Spectrum

In this section we consider the case when the operators $L(t) \colon C \to \mathbb{R}^n$ are defined only for $t \ge 0$ and we describe all possible forms of the corresponding one-sided spectrum.

Definition 10.2. The *one-sided spectrum* of Eq. (2.12) is the set Σ^+ of all numbers $a \in \mathbb{R}$ such that the evolution family

$$T_a(t,s) = e^{-a(t-s)}T(t,s)$$

has no exponential dichotomy on $\mathbb{R}_0^+$.

The following result describes all possible forms of the one-sided spectrum. We consider the same numbers c_m as in Sec. 10.2. Note that the spaces $E_{c_m}(s)$ are independent of the choice of these numbers.

Theorem 10.3 (One-sided spectrum). *Assume that Σ is nonempty. For the one-sided spectrum of Eq. (2.12), one of the alternatives in Theorem 10.1 holds. Moreover, for each $s \geq 0$ and $\varphi \in E_{c_{m-1}}(s) \setminus E_{c_m}(s)$ we have*

$$\left(\liminf_{t \to +\infty} \frac{1}{t} \log \|T(t,s)\varphi\|, \limsup_{t \to +\infty} \frac{1}{t} \log \|T(t,s)\varphi\| \right) \subset (a_m, b_m).$$

Proof. Given $a \in \mathbb{R} \setminus \Sigma^+$, we denote by d_a the dimension of some unstable space of the exponential dichotomy for the evolution family $T_a(t,s)$. The first statement can be obtained as in the proof of Theorem 10.1 replacing $\dim F_a(s)$ by d_a.

For the second statement, since $c_{m-1} \notin \Sigma^+$, the evolution family $T_{c_{m-1}}(t,s)$ has an exponential dichotomy and so there exists projections $P(t)$ for $t \geq 0$ satisfying (3.2) and there exist constants $\lambda, D > 0$ such that

$$\|T(t,s)P(s)\| \leq De^{(c_{m-1}-\lambda)(t-s)} \tag{10.17}$$

and

$$\|\overline{T}(s,t)Q(t)\| \leq De^{-(\lambda+c_{m-1})(t-s)}$$

for $t \geq s$, where $Q(t) = \mathrm{Id} - P(t)$. Moreover, $\operatorname{Im} P(t) = E_{c_{m-1}}(t)$ for $t \geq 0$. Hence, for $\varphi \in E_{c_{m-1}}(s)$, it follows from (10.17) that

$$\limsup_{t \to +\infty} \frac{1}{t} \log \|T(t,s)\varphi\| \leq c_{m-1} - \lambda < c_{m-1}$$

and letting $c_{m-1} \searrow b_m$ we obtain

$$\limsup_{t \to +\infty} \frac{1}{t} \log \|T(t,s)\varphi\| \leq b_m.$$

Similarly, since $c_m \notin \Sigma^+$, there exist projections $\overline{P}(t)$ for $t \geq 0$ satisfying (3.2) and there exist constant $\mu, D > 0$ such that

$$\|T(t,s)\overline{P}(s)\| \leq De^{(c_m-\mu)(t-s)}$$

and

$$\|\overline{T}(s,t)\overline{Q}(t)\| \leq De^{-(\mu+c_m)(t-s)}$$

for $t \geq s$, where $\overline{Q}(t) = \mathrm{Id} - \overline{P}(t)$. Now take $\varphi \in E_{c_{m-1}}(s) \setminus E_{c_m}(s)$ and write it in the form $\varphi = x + y$ with $x \in E_{c_m}(s)$ and $y \in F_{c_m}(s) \setminus \{0\}$. We have

$$\|y\| \leq De^{-(\mu+c_m)(t-s)}\|\overline{T}(t,s)y\| \quad \text{for } t \geq s$$

and so

$$\|\overline{T}(t,s)\varphi\| \geq \|\overline{T}(t,s)y\| - \|\overline{T}(t,s)x\|$$

$$\geq \frac{1}{D} e^{(\mu+c_m)(t-s)} \|y\| - D e^{(c_m-\mu)(t-s)} \|x\|.$$

Therefore,

$$\liminf_{t\to+\infty} \frac{1}{t} \log\|T(t,s)\varphi\| \geq \mu + c_m > c_m$$

and letting $c_m \nearrow a_m$ yields the inequality

$$\liminf_{t\to+\infty} \frac{1}{t} \log\|T(t,s)\varphi\| \geq a_m.$$

This completes the proof of the theorem. $\qquad\square$

10.4 Persistence of the Spectrum

In this section we show that the asymptotic behavior described in Theorem 10.2 persists under sufficiently small perturbations. An entirely analogous statement holds for the one-sided asymptotic behavior described in Theorem 10.3 and so we omit the details.

More precisely, we consider the equation

$$v' = L(t)v_t + g(t, v_t), \qquad (10.18)$$

with the standing assumptions of Sec. 2.2 and where $g\colon \mathbb{R} \times C \to \mathbb{R}^n$ is a measurable function. We continue to assume that $r > 0$.

We start with a preliminary result.

Proposition 10.5. *Any solution* $v\colon [s-r, a) \to \mathbb{R}^n$ *of Eq. (10.18) satisfying*

$$\lim_{t\to+\infty} \int_t^{t+1} \frac{|g(\tau, v_\tau)|}{\|v_\tau\|} \, d\tau = 0 \qquad (10.19)$$

can be continued to the interval $[s - r, +\infty)$.

Proof. We define a measurable function $\gamma\colon \mathbb{R} \to [0, +\infty]$ by

$$\gamma(t) = \frac{|g(t, v_t)|}{\|v_t\|}.$$

In view of (10.19), we have

$$\lim_{t\to+\infty} \int_t^{t+1} \gamma(\tau) \, d\tau = 0. \qquad (10.20)$$

Proceeding as in (2.14) and (2.15) with $L(\tau)v_\tau$ replaced by $L(\tau)v_\tau + g(\tau, v_\tau)$, we obtain

$$\|v_t\| \leq \|v_s\| + \int_s^t \left(\Pi(\tau)\|v_\tau\| + |g(\tau, v_\tau)| \right) d\tau$$

$$\leq \|v_s\| + \int_s^t \left(\Pi(\tau) + \gamma(\tau) \right) \|v_\tau\| \, d\tau$$

for $t \in [s, a)$. Applying Gronwall's lemma we conclude that

$$\|v_t\| \leq \|v_s\| \exp\left(\int_s^t \left(\Pi(\tau) + \gamma(\tau) \right) d\tau \right)$$

for $t \in [s, a)$ and so also

$$|L(t)v_t + g(t, v_t)| \leq \left(\Pi(t) + \gamma(t) \right) \|v_t\|$$

$$\leq \left(\Pi(t) + \gamma(t) \right) \|v_s\| \exp\left(\int_s^a \left(\Pi(\tau) + \gamma(\tau) \right) d\tau \right)$$

for $t \in [s, a)$. Since Π and γ are locally integrable (using also (10.19)), one can show as in (2.17) that v is uniformly continuous on $[s - r, a)$. One can now proceed as in the proof of Theorem 2.2 to verify that

$$A = \{(t, v_t) : t \in [s, a)\}$$

is contained in a compact set. Hence, it follows from Theorem 2.3 that the solution v is continuable. $\qquad\square$

Finally, we show that the asymptotic behavior of the linear equation persists under sufficiently small perturbations.

Theorem 10.4. *Assume that condition (2.22) holds and that the spectrum Σ is nonempty. Let $v \colon [s - r, +\infty) \to \mathbb{R}^n$ be a solution of Eq. (10.18) satisfying (10.19). If there exists $p \in \mathbb{N}$ such that*

$$\liminf_{t \to +\infty} \frac{1}{t} \log\|v_t\| \geq a_p,$$

then there exists $i \in \{1, \ldots, p\}$ such that

$$\left(\liminf_{t \to +\infty} \frac{1}{t} \log\|v_t\|, \limsup_{t \to +\infty} \frac{1}{t} \log\|v_t\| \right) \subset (a_i, b_i).$$

Proof. We first establish an auxiliary result.

Lemma 10.4. *Given $c > 0$, there exists $N > 0$ such that*

$$\|v_t\| \leq N\|v_s\| e^{c(t-s)} e^{N \int_s^t \gamma(\tau) \, d\tau}$$

for all $t \geq s$.

Proof of the lemma. By Proposition 10.3, we have $\Sigma \subset (-\infty, \omega]$, with ω as in (2.21). In particular, Σ is bounded from above and so $\sup \Sigma = b_1$. Now take $c > b_1$. Since $b_1 = \sup \Sigma$, there exist constants $\lambda, D > 0$ such that

$$\|T(t,s)\| \le De^{(c-\lambda)(t-s)} \quad \text{for } t \ge s$$

and so, by Proposition 3.1,

$$\|v_t\| \le De^{(c-\lambda)(t-s)}\|v_s\| + N \int_s^t e^{(c-\lambda)(t-\tau)}\gamma(\tau)\|v_\tau\|\, d\tau$$

for all $t \ge s$ and some constant $N \ge D$. Therefore,

$$e^{-(c-\lambda)(t-s)}\|v_s\| \le N\|v_s\| + N \int_s^t e^{-(c-\lambda)(\tau-s)}\gamma(\tau)\|v_\tau\|\, d\tau$$

for $t \ge s$. Applying Gronwall's lemma, we obtain

$$e^{-(c-\lambda)(t-s)}\|v_t\| \le N\|v_s\|e^{N\int_s^t \gamma(\tau)\, d\tau}$$

and hence,

$$\|v_t\| \le N\|v_s\|e^{(c-\lambda)(t-s)}e^{N\int_s^t \gamma(\tau)\, d\tau} \le N\|v_s\|e^{c(t-s)}e^{N\int_s^t \gamma(\tau)\, d\tau}.$$

This concludes the proof of the lemma. $\qquad\square$

Now take $m \in \mathbb{N}$ such that

$$D < e^{\lambda m/2}. \tag{10.21}$$

It follows from Lemma 10.4 that for $km \le t \le (k+1)m$ with $k \in \mathbb{N}$, we have

$$C^{-1}e^{-c((k+1)m-t)}\|v_{(k+1)m}\| \le \|v_t\| \le Ce^{c(t-km)}\|v_{km}\|, \tag{10.22}$$

where $C = Ne^{N\Gamma}$ and

$$\Gamma = \sup_{s \ge 0} \int_s^{s+1} \gamma(\tau)\, d\tau \le m \sup_{s \ge 0} \int_s^{s+1} \gamma(\tau)\, d\tau < +\infty,$$

using property (10.20).

Given $d \in \mathbb{R} \setminus \Sigma$, the evolution family $T_d(t,s) = e^{-d(t-s)}T(t,s)$ has an exponential dichotomy on $\mathbb{R}$ and so there exist projections $P(t)$ satisfying (3.2) and there exist constants $\lambda, D > 0$ such that

$$\|T(t,s)P(s)\| \le De^{(d-\lambda)(t-s)} \tag{10.23}$$

and

$$\|\overline{T}(s,t)Q(t)\| \le De^{-(\lambda+d)(t-s)} \tag{10.24}$$

for $t \geq s$, where $Q(t) = \mathrm{Id} - P(t)$. We write $v_t = x_t + y_t$, where

$$x_t = P(t)v_t \quad \text{and} \quad y_t = Q(t)v_t.$$

By Theorem 3.6, for $t \geq km$ we have

$$x_t = T(t, km)x_{km} + \int_{km}^{t} T(t, \tau)P_0(\tau)g(\tau, v_\tau)\, d\tau \qquad (10.25)$$

and

$$y_t = T(t, km)y_{km} + \int_{km}^{t} T(t, \tau)Q_0(\tau)g(\tau, v_\tau)\, d\tau. \qquad (10.26)$$

By Proposition 3.1, (10.24) and (10.26), for $t \geq km$ we obtain

$$\|y_t\| \geq \|T(t, km)y_{km}\| - \left\| \int_{km}^{t} T(t, \tau)Q_0(\tau)g(\tau, v_\tau)\, d\tau \right\|$$

$$\geq D^{-1}e^{(d+\lambda)(t-km)}\|y_{km}\| - NK \int_{km}^{t} e^{c(t-\tau)}\gamma(\tau)\|v_\tau\|\, d\tau.$$

Hence, by (10.22), for $km \leq t \leq (k+1)m$ we have

$$\|y_t\| \geq D^{-1}e^{(d+\lambda)(t-km)}\|y_{km}\|$$

$$- NKC\|v_{km}\| \int_{km}^{(k+1)m} e^{c(t-\tau)}\gamma(\tau)e^{c(\tau-km)}\, d\tau$$

$$\geq D^{-1}e^{(d+\lambda)(t-km)}\|y_{km}\| - \rho\Gamma_k\|v_{km}\|$$

for some constant $\rho = \rho(m) > 0$, where

$$\Gamma_k = \int_{km}^{(k+1)m} \gamma(\tau)\, d\tau.$$

Similarly, by Proposition 3.1, (10.23) and (10.25), for $km \leq t \leq (k+1)m$ we have

$$\|x_t\| \leq De^{(d-\lambda)(t-km)}\|x_{km}\| + NK \int_{km}^{t} e^{c(t-\tau)}\gamma(\tau)\|v_\tau\|\, d\tau$$

$$\leq De^{(d-\lambda)(t-km)}\|x_{km}\|$$

$$+ NKC\|v_{km}\| \int_{km}^{(k+1)m} e^{c(t-\tau)}e^{c(\tau-km)}\gamma(\tau)\, d\tau$$

$$\leq De^{(d-\lambda)(t-km)}\|x_{km}\| + \rho\Gamma_k\|v_{km}\|.$$

Therefore, using (10.21), we obtain

$$\|x_t\| \leq De^{(d-\lambda)(t-km)}\|x_{km}\| + \rho\Gamma_k\big(\|x_{km}\| + \|y_{km}\|\big)$$

$$\leq e^{(d-\lambda/2)(t-km)}\|x_{km}\| + \rho\Gamma_k\big(\|x_{km}\| + \|y_{km}\|\big) \qquad (10.27)$$

and

$$\|y_t\| \geq D^{-1} e^{(d+\lambda)(t-km)} \|y_{km}\| - \rho\Gamma_k \big(\|x_{km}\| + \|y_{km}\|\big)$$
$$\geq e^{(d+\lambda/2)(t-km)} \|y_{km}\| - \rho\Gamma_k \big(\|x_{km}\| + \|y_{km}\|\big). \tag{10.28}$$

Lemma 10.5. *Either*

$$\|y_{km}\| \leq \|x_{km}\| \quad \text{for any sufficiently large } k \tag{10.29}$$

or

$$\|x_{km}\| \leq \|y_{km}\| \quad \text{for any sufficiently large } k. \tag{10.30}$$

Proof of the lemma. Assume that (10.29) does not hold. Then there exists k_0 arbitrarily large such that

$$\|x_{k_0 m}\| < \|y_{k_0 m}\|.$$

We prove by induction that if k_0 is sufficiently large, then $\|x_{km}\| < \|y_{km}\|$ for all $k \geq k_0$. Assume that $\|x_{km}\| < \|y_{km}\|$ for some $k \geq k_0$. By (10.27) and (10.28), we have

$$\|x_{(k+1)m}\| \leq (e^{(d-\lambda/2)m} + \rho\Gamma_k)\|x_{km}\| + \rho\Gamma_k\|y_{km}\|$$

and

$$\|y_{(k+1)m}\| \geq (e^{(d+\lambda/2)m} - \rho\Gamma_k)\|y_{km}\| - \rho\Gamma_k\|x_{km}\|.$$

Therefore,

$$\|x_{(k+1)m}\| \leq (e^{(d-\lambda/2)m} + 2\rho\Gamma_k)\|y_{km}\|$$

and

$$\|y_{(k+1)m}\| \geq (e^{(d+\lambda/2)m} - 2\rho\Gamma_k)\|y_{km}\|$$

(provided that k_0 is sufficiently large), which yields the inequality

$$\|x_{(k+1)m}\| \leq \frac{e^{(d-\lambda/2)m} + 2\rho\Gamma_k}{e^{(d+\lambda/2)m} - 2\rho\Gamma_k} \|y_{(k+1)m}\|.$$

On the other hand, by (10.20) we have

$$\lim_{k\to\infty} \frac{e^{(d-\lambda/2)m} + 2\rho\Gamma_k}{e^{(d+\lambda/2)m} - 2\rho\Gamma_k} < 1.$$

Hence,

$$\|x_{(k+1)m}\| < \|y_{(k+1)m}\| \quad \text{for } k \geq k_0,$$

provided that k_0 is sufficiently large. This shows that if (10.29) fails, then (10.30) holds and the proof of the lemma is complete. $\square$

Lemma 10.6. *One of the following alternatives holds:*

(1)

$$\limsup_{t \to +\infty} \frac{1}{t} \log \|v_t\| < d \qquad (10.31)$$

and

$$\lim_{k \to +\infty} \frac{\|y_{km}\|}{\|x_{km}\|} = 0; \qquad (10.32)$$

(2)

$$\liminf_{t \to +\infty} \frac{1}{t} \log \|v_t\| > d$$

and

$$\lim_{k \to +\infty} \frac{\|x_{km}\|}{\|y_{km}\|} = 0. \qquad (10.33)$$

Proof of the lemma. Assume that (10.29) holds and let

$$S = \limsup_{k \to \infty} \frac{\|y_{km}\|}{\|x_{km}\|}.$$

By (10.29) we have $0 \le S \le 1$. Moreover, by (10.27) with $t = (k+1)m$ we obtain

$$\|x_{(k+1)m}\| \le (e^{(d-\lambda/2)m} + 2\rho\Gamma_k)\|x_{km}\| \qquad (10.34)$$

for all large k. Hence, it follows from (10.28) with $t = (k+1)m$ that

$$\frac{\|y_{(k+1)m}\|}{\|x_{(k+1)m}\|} \ge \frac{e^{(d+\lambda/2)m} - \rho\Gamma_k}{e^{(d-\lambda/2)m} + 2\rho\Gamma_k} \cdot \frac{\|y_{km}\|}{\|x_{km}\|} - \frac{\rho\Gamma_k}{e^{(d-\lambda/2)m} + 2\rho\Gamma_k}. \qquad (10.35)$$

Taking limits when $k \to \infty$ we obtain

$$\delta := \lim_{k \to \infty} \frac{e^{(d+\lambda/2)m} - \rho\Gamma_k}{e^{(d-\lambda/2)m} + 2\rho\Gamma_k} > 1$$

and

$$\lim_{k \to \infty} \frac{\rho\Gamma_k}{e^{(d-\lambda/2)m} + 2\rho\Gamma_k} = 0.$$

Hence, letting $k \to \infty$ in (10.35), we find that $S \ge \delta S$. This implies that $S = 0$ and so (10.32) holds. In order to establish property (10.31), take k_0 such that (10.34) holds for all $k \ge k_0$. Then

$$\|x_{km}\| \le \|x_{k_0 m}\| \prod_{j=k_0}^{k-1} \left(e^{(d-\lambda/2)m} + 2\rho\Gamma_j\right)$$

$$= e^{(d-\lambda/2)m(k-k_0)} \prod_{j=k_0}^{k-1} \left(1 + 2\rho e^{-(d-\lambda/2)m}\Gamma_j\right)\|x_{k_0 m}\|.$$

Therefore, by (10.22), for $k \geq k_0$ and $km \leq t \leq (k+1)m$ we have

$$\|v_t\| \leq Ce^{cm}\|v_{km}\| \leq Ce^{cm}(\|x_{km}\| + \|y_{km}\|) \leq 2Ce^{cm}\|x_{km}\|$$

$$\leq 2Ce^{cm}e^{(d-\lambda/2)m(k-k_0)} \prod_{j=k_0}^{k-1} \left(1 + 2\rho e^{-(d-\lambda/2)m}\Gamma_j\right)\|x_{k_0 m}\|.$$

Note that for $t \geq km \geq k_0 m$,

$$\|v_t\| \leq 2Ce^{cm}e^{(d-\lambda/2)t}e^{-(d-\lambda/2)mk_0}$$

$$\times \prod_{j=k_0}^{k-1} \left(1 + 2\rho e^{-(d-\lambda/2)m}\Gamma_j\right)\|x_{k_0 m}\|. \tag{10.36}$$

On the other hand, by (10.20) we have

$$\frac{1}{km}\log \prod_{j=k_0}^{k-1} \left(1 + 2\rho e^{-(d-\lambda/2)m}\Gamma_j\right) = \frac{1}{km}\sum_{j=k_0}^{k-1} \log\left(1 + 2\rho e^{-(d-\lambda/2)m}\Gamma_j\right)$$

$$\leq \frac{1}{km}2\rho e^{-(d-\lambda/2)m} \sum_{j=k_0}^{k-1} \Gamma_j \to 0$$

when $k \to \infty$. Hence, letting $t \to +\infty$ in (10.36) we obtain

$$\limsup_{t \to +\infty} \frac{1}{t}\log\|v_t\| \leq d - \frac{\lambda}{2} < d,$$

which yields property (10.31).

Now assume that (10.30) holds and let

$$S = \limsup_{k \to \infty} \frac{\|x_{km}\|}{\|y_{km}\|}.$$

By (10.30) we have $0 \leq S \leq 1$. Moreover, by (10.28) with $t = (k+1)m$ we obtain

$$\|y_{(k+1)m}\| \geq (e^{(d+\lambda/2)m} - 2\rho\Gamma_k)\|y_{km}\| \tag{10.37}$$

for all large k. Hence, it follows from (10.27) that

$$\frac{\|x_{(k+1)m}\|}{\|y_{(k+1)m}\|} \leq \frac{e^{(d-\lambda/2)m} + \rho\Gamma_k}{e^{(d+\lambda/2)m} - 2\rho\Gamma_k} \cdot \frac{\|x_{km}\|}{\|y_{km}\|} + \frac{\rho\Gamma_k}{e^{(d+\lambda/2)m} - 2\rho\Gamma_k} \tag{10.38}$$

for all large k. Taking limits when $k \to \infty$ we obtain

$$\delta := \lim_{k \to \infty} \frac{e^{(d-\lambda/2)m} + \rho\Gamma_k}{e^{(d+\lambda/2)m} - 2\rho\Gamma_k} < 1$$

and

$$\lim_{k \to \infty} \frac{\rho\Gamma_k}{e^{(d+\lambda/2)m} - 2\rho\Gamma_k} = 0.$$

Hence, letting $k \to \infty$ in (10.38), we find that $S \leq \delta S$. This implies that $S = 0$ and so (10.33) holds. Now take k_0 such that (10.37) holds for all $k \geq k_0$. Then

$$\|y_{(k+1)m}\| \geq \|y_{k_0 m}\| e^{(d+\lambda/2)m(k-k_0)} \prod_{j=k_0}^{k} \left(1 - 2\rho e^{-(d+\lambda/2)m} \Gamma_j\right).$$

By (10.22), for $k \geq k_0$ and $km \leq t \leq (k+1)m$ we have

$$\|v_t\| \geq C^{-1} \|v_{(k+1)m}\| \geq C^{-1} \|y_{(k+1)m}\|$$

$$\geq C^{-1} e^{-cm} e^{(d+\lambda/2)m(k-k_0)} \prod_{j=k_0}^{k} \left(1 - 2\rho e^{-(d+\lambda/2)m} \Gamma_j\right) \|y_{k_0 m}\|.$$

Note that for $(k+1)m \geq t \geq km \geq k_0 m$,

$$\|v_t\| \geq C^{-1} e^{-(c+d+\lambda/2)m} e^{(d+\lambda/2)t} e^{-(d+\lambda/2)mk_0}$$
$$\times \prod_{j=k_0}^{k} \left(1 - 2\rho e^{-(d+\lambda/2)m} \Gamma_j\right) \|x_{k_0 m}\|. \tag{10.39}$$

On the other hand, by (10.20), for k_0 sufficiently large we have

$$\frac{1}{km} \log \prod_{j=k_0}^{k} \left(1 - 2\rho e^{-(d+\lambda/2)m} \Gamma_j\right) = \frac{1}{km} \sum_{j=k_0}^{k} \log\left(1 - 2\rho e^{-(d+\lambda/2)m} \Gamma_j\right)$$

$$\geq -\frac{1}{km} 4\rho e^{-(d+\lambda/2)m} \sum_{j=k_0}^{k} \Gamma_j$$

and so

$$\frac{1}{km} \log \prod_{j=k_0}^{k} \left(1 - 2\rho e^{-(d+\lambda/2)m} \Gamma_j\right) \to 0$$

when $k \to \infty$. Hence, letting $t \to +\infty$ in (10.39) we obtain

$$\liminf_{t \to +\infty} \frac{1}{t} \log\|v_t\| \geq d + \frac{\lambda}{2} > d.$$

This completes the proof of the lemma. $\qquad\square$

We proceed with the proof of the theorem. Let v be a solution as in the statement of the theorem. First note that given $c > b_1$, it follows from Lemma 10.4 and (10.20) that

$$\limsup_{t \to +\infty} \frac{1}{t} \log\|v_t\| \leq c + \limsup_{t \to +\infty} \frac{N}{t} \int_s^t \gamma(\tau)\, d\tau = c.$$

Letting $\varepsilon \to 0$ we obtain

$$\limsup_{t \to +\infty} \frac{1}{t} \log \|v_t\| \le b_1. \tag{10.40}$$

Take $c_i \in (b_{i+1}, a_i)$ for $i \in \{1, \dots, p\}$. By Lemma 10.6, for each such i either

$$\limsup_{t \to +\infty} \frac{1}{t} \log \|v_t\| < c_i$$

or

$$\liminf_{t \to +\infty} \frac{1}{t} \log \|v_t\| > c_i.$$

Together with (10.40), this implies that there exists $i \in \{1, \dots, p\}$ such that

$$\limsup_{t \to +\infty} \frac{1}{t} \log \|v_t\| < c_{i-1}$$

and

$$\liminf_{t \to +\infty} \frac{1}{t} \log \|v_t\| > c_i.$$

Letting $c_{i-1} \searrow b_i$ and $c_i \nearrow a_i$, we finally obtain

$$a_i \le \liminf_{t \to +\infty} \frac{1}{t} \log \|v_t\| \le \limsup_{t \to +\infty} \frac{1}{t} \log \|v_t\| \le b_i.$$

This completes the proof of the theorem. $\qquad\square$

We also consider the particular case of linear perturbations. Consider the linear equation

$$v' = L(t)v_t + M(t)v_t \tag{10.41}$$

for some linear operators $L(t), M(t) \colon C \to \mathbb{R}^n$, for $t \in \mathbb{R}$, satisfying the standing assumptions of Sec. 4.1. The following result is an immediate consequence of Theorem 10.4.

Theorem 10.5. *Assume that* $\|M(t)\| \le \gamma(t)$ *for some continuous function* $\gamma \colon \mathbb{R} \to \mathbb{R}_0^+$ *satisfying* (10.20). *Then the equations* $v' = L(t)v_t$ *and* (10.41) *have the same spectra.*

Bibliography

[1] L. Barreira, D. Dragičević and C. Valls, Admissibility on the half line for evolution families, *J. Anal. Math.* **132**, pp. 157–176 (2017).

[2] L. Barreira, D. Dragičević and C. Valls, *Admissibility and Hyperbolicity, SpringerBriefs in Mathematics.* Springer, Cham (2018).

[3] L. Barreira and Ya. Pesin, *Lyapunov Exponents and Smooth Ergodic Theory, University Lecture Series*, Vol. 23, American Mathematical Society, Providence, RI (2002).

[4] L. Barreira and Ya. Pesin, *Nonuniform Hyperbolicity. Dynamics of Systems with Nonzero Lyapunov Exponents, Encyclopedia of Mathematics and its Applications*, Vol. 115. Cambridge University Press, Cambridge (2007).

[5] L. Barreira and C. Valls, Robustness of nonuniform exponential dichotomies in Banach spaces, *J. Differential Equations* **244**, pp. 2407–2447 (2008).

[6] L. Barreira and C. Valls, A simple proof of the Grobman–Hartman theorem for nonuniformly hyperbolic dynamics, *Nonlinear Anal.* **74**, pp. 7210–7225 (2011).

[7] L. Barreira and C. Valls, Smooth robustness of exponential dichotomies, *Proc. Amer. Math. Soc.* **139**, pp. 999–1012 (2011).

[8] L. Barreira and C. Valls, A Perron-type theorem for nonautonomous differential equations, *J. Differential Equations* **258**, pp. 339–361 (2015).

[9] L. Barreira and C. Valls, Evolution families and nonuniform spectrum, *Electron. J. Qual. Theory Differ. Equ.* **2016**, Paper No. 48, 13 pp. (2016).

[10] L. Barreira and C. Valls, Robustness of hyperbolicity in delay equations, *J. Dynam. Differential Equations* **32**, pp. 1–22 (2020).

[11] L. Barreira and C. Valls, Center manifolds for delay equations, *Funkcial. Ekvac.* (to appear).

[12] L. Barreira and C. Valls, Hyperbolicity via admissibility in delay equations, *Indiana Univ. Math. J.* (to appear).

[13] G. Belickiĭ, Functional equations, and conjugacy of local diffeomorphisms of finite smoothness class, *Functional Anal. Appl.* **7**, pp. 268–277 (1973).

[14] G. Belickiĭ, Equivalence and normal forms of germs of smooth mappings, *Russian Math. Surveys* **33**, pp. 107–177 (1978).

[15] R. Bellman and K. Cooke, *Differential-Difference Equations*. Academic Press, New York-London (1963).

[16] R. Bellman and J. Danskin, A survey of the mathematical theory of time lag, retarded control, and hereditary processes, Report 256. The RAND Corporation, Santa Monica, CA (1954).

[17] N. Bhatia and G. Szegö, *Stability Theory of Dynamical Systems, Grundlehren der mathematischen Wissenschaften*, Vol. 161. Springer-Verlag, New York-Berlin (1970).

[18] C. Blázquez, Transverse homoclinic orbits in periodically perturbed parabolic equations, *Nonlinear Anal.* **10**, pp. 1277–1291 (1986).

[19] J. Carr, *Applications of Centre Manifold Theory, Applied Mathematical Sciences*, Vol. 35. Springer-Verlag, New York-Berlin (1981).

[20] N. Chafee, The bifurcation of one or more closed orbits from an equilibrium point of an autonomous differential equation, *J. Differential Equations* **4**, pp. 661–679 (1968).

[21] N. Chafee, A bifurcation problem for functional differential equations of finitely retarded type, *J. Math. Anal. Appl.* **35**, pp. 312–348 (1971).

[22] C. Chicone and Yu. Latushkin, Center manifolds for infinite dimensional nonautonomous differential equations, *J. Differential Equations* **141**, pp. 356–399 (1997).

[23] C. Chicone and Yu. Latushkin, *Evolution Semigroups in Dynamical Systems and Differential Equations, Mathematical Surveys and Monographs*, Vol. 70. American Mathematical Society, Providence, RI (1999).

[24] S.-N. Chow and H. Leiva, Dynamical spectrum for time dependent linear systems in Banach spaces, *Jap. J. Ind. Appl. Math.* **11**, pp. 379–415 (1994).

[25] S.-N. Chow and H. Leiva, *Dynamical spectrum for skew product flow in Banach spaces, Boundary Value Problems for Functional-Differential Equations*. World Scientific Publishing, River Edge, NJ, pp. 85–105 (1995).

[26] S.-N. Chow and H. Leiva, Existence and roughness of the exponential dichotomy for skew-product semiflow in Banach spaces, *J. Differential Equations* **120**, pp. 429–477 (1995).

[27] S.-N. Chow and K. Lu, C^k centre unstable manifolds, *Proc. Roy. Soc. Edinburgh Sect. A* **108**, pp. 303–320 (1988).

[28] C. Coffman, Asymptotic behavior of solutions of ordinary difference equations, *Trans. Amer. Math. Soc.* **110**, pp. 22–51 (1964).

[29] C. Constantine and T. Savits, A multivariate Faà di Bruno formula with applications, *Trans. Amer. Math. Soc.* **348**, pp. 503–520 (1996).

[30] W. Coppel, *Stability and Asymptotic Behavior of Differential Equations*. D. C. Heath and Co., Boston, Mass. (1965).

[31] W. Coppel, Dichotomies and reducibility, *J. Differential Equations* **3**, pp. 500–521 (1967).

[32] W. Coppel, *Dichotomies in Stability Theory, Lecture Notes in Mathematics*, Vol. 629. Springer-Verlag, Berlin-New York (1978).

[33] Ju. Dalec'kiĭ and M. Kreĭn, *Stability of Solutions of Differential Equations in Banach Space, Translations of Mathematical Monographs*, Vol. 43. American Mathematical Society, Providence, RI (1974).

[34] O. Diekmann, S. van Gils, S. Verduyn-Lunel and H.-O. Walther, *Delay Equations. Functional, Complex, and Nonlinear Analysis, Applied Mathematical Sciences*, Vol. 110. Springer-Verlag, New York (1995).

[35] T. Faria and L. Magalhães, Normal forms for retarded functional differential equations with parameters and applications to Bogdanov–Takens singularity, *J. Differential Equations* **122**, pp. 201–224 (1995).

[36] T. Faria and L. Magalhães, Normal forms for retarded functional differential equations with parameters and applications to Hopf bifurcation, *J. Differential Equations* **122**, pp. 181–200 (1995).

[37] G. Farkas, A Hartman–Grobman result for retarded functional differential equations with an application to the numerics around hyperbolic equilibria, *Z. Angew. Math. Phys.* **52**, pp. 421–432 (2001).

[38] V. Fodcuk, Integral varieties for nonlinear differential equations with retarded arguments (in Russian), *Ukrain. Mat. Z.* **21**, pp. 627–639 (1969).

[39] V. Fodcuk, Integral manifolds for nonlinear differential equations with retarded arguments (in Russian), *Differencial'nye Uravnenija* **6**, pp. 798–808 (1970).

[40] D. Gilsinn, Estimating critical Hopf bifurcation parameters for a second order delay differential equation with application to machine tool chatter, *Nonlinear Dynam.* **30**, pp. 103–154 (2002).

[41] D. Grobman, Homeomorphism of systems of differential equations, *Dokl. Akad. Nauk SSSR* **128**, pp. 880–881 (1959).

[42] D. Grobman, Topological classification of neighborhoods of a singularity in *n*-space, *Mat. Sb. (N.S.)* **56 (98)**, pp. 77–94 (1962).

[43] J. Guckenheimer and P. Holmes, *Nonlinear Oscillations: Dynamical System and Bifurcations of Vector Fields, Applied Mathematical Sciences*, Vo l. 42. Springer-Verlag, New York (1983).

[44] S. Guo and L. Huang, Hopf bifurcating periodic orbits in a ring of neurons with delays, *Physica D* **183**, pp. 19–44 (2003).

[45] S. Guo, L. Huang and L. Wang, Linear stability and Hopf bifurcation in a two-neuron network with three delays, *Internat. J. Bifur. Chaos Appl. Sci. Engrg.* **14**, pp. 2799–2810 (2004).

[46] W. Hahn, *Stability of Motion, Die Grundlehren der mathematischen Wissenschaften*, Vol. 138. Springer-Verlag New York, Inc., New York (1967).

[47] A. Halanay, *Differential Equations: Stability, Oscillations, Time Lags*. Academic Press, New York-London (1966).

[48] J. Hale, Linear functional-differential equations with constant coefficients, *Contrib. Differ. Equations* **2**, pp. 291–317 (1963).

[49] J. Hale, *Functional Differential Equations, Applied Mathematical Sciences*, Vol. 3. Springer-Verlag, New York-Heidelberg (1971).

[50] J. Hale, *Theory of Functional Differential Equations, Applied Mathematical Sciences*, Vol. 3, Springer-Verlag, New York-Heidelberg (1977).

[51] J. Hale, *Asymptotic Behavior of Dissipative Systems, Mathematical Surveys and Monographs*, Vol. 25. American Mathematical Society, Providence, RI (1988).

[52] J. Hale, L. Magalhães and W. Oliva, *Dynamics in Infinite Dimensions*, with an appendix by K. Rybakowski, *Applied Mathematical Sciences*, Vol. 47. Springer-Verlag, New York (2002).

[53] J. Hale and C. Perelló, The neighborhood of a singular point of functional differential equations, *Contrib. Differ. Equations* **3**, pp. 351–375 (1964).

[54] J. Hale and S. Verduyn Lunel, *Introduction to Functional-Differential Equations, Applied Mathematical Sciences*, Vol. 99. Springer-Verlag, New York (1993).

[55] P. Hartman, A lemma in the theory of structural stability of differential equations, *Proc. Amer. Math. Soc.* **11**, pp. 610–620 (1960).

[56] P. Hartman, On the local linearization of differential equations, *Proc. Amer. Math. Soc.* **14**, pp. 568–573 (1963).

[57] P. Hartman and A. Wintner, Asymptotic integrations of linear differential equations, *Amer. J. Math.* **77**, pp. 45–86 (1955).

[58] D. Henry, *Geometric Theory of Semilinear Parabolic Equations, Lecture Notes in Mathematics*, Vol. 840. Springer-Verlag, Berlin-New York (1981).

[59] D. Henry, *Variedades invariantes perto dum ponto fixo*, Lecture Notes Univ. São Paulo, Brazil (1983).

[60] E. Hille and R. Phillips, *Functional Analysis and Semi-Groups, Colloquium Publications*, Vol. 31. American Mathematical Society, Providence, RI (1957).

[61] Y. Hino, S. Murakami and T. Naito, *Functional-Differential Equations with Infinite Delay, Lecture Notes in Mathematics*, Vol. 1473. Springer-Verlag, Berlin (1991).

[62] M. Hirsch, C. Pugh and M. Shub, *Invariant Manifolds, Lecture Notes in Mathematics*, Vol. 583. Springer-Verlag, Berlin-New York (1977).

[63] N. Huy, Exponential dichotomy of evolution equations and admissibility of function spaces on a half-line, *J. Funct. Anal.* **235**, pp. 330–354 (2006).

[64] R. Johnson, K. Palmer and G. Sell, Ergodic properties of linear dynamical systems, *SIAM J. Math. Anal.* **18**, pp. 1–33 (1987).

[65] R. Johnson and G. Sell, Smoothness of spectral subbundles and reducibility of quasiperiodic linear differential systems, *J. Differential Equations* **41**, pp. 262–288 (1981).

[66] T. Kalmár-Nagy, G. Stépán and F. Moon, Subcritical Hopf bifurcation in the delay equation model for machine tool vibrations, *Nonlinear Dynam.* **26**, pp. 121–142 (2001).

[67] A. Kelley, The stable, center-stable, center, center-unstable and unstable manifolds, *J. Differential Equations* **3**, pp. 546–570 (1967).

[68] Y. Kōmura, Nonlinear semi-groups in Hilbert space, *J. Math. Soc. Japan* **19**, pp. 493–507 (1967).

[69] N. Krasovskiĭ, On the theory of the second method of A. M. Lyapunov for the investigation of stability (in Russian), *Mat. Sb. N.S.* **40**, pp. 57–64 (1956).

[70] N. Krasovskiĭ, *Stability of Motion. Applications of Lyapunov's Second Method to Differential Systems and Equations with Delay* (transl. from the Russian, 1959). Stanford University Press, Stanford, CA (1963).

[71] J. Kurzweil, *Invariant manifolds for flows, Differential Equations and Dynamical Systems (Proc. Internat. Sympos., Mayaguez, PR, 1965)*. Academic Press, New York, pp. 431–468 (1967).

[72] J. Kurzweil, Invariant manifolds I, *Comm. Math. Univ. Carolinae* **11**, pp. 309–336 (1970).

[73] J. Lamb and J. Roberts, Time-reversal symmetry in dynamical systems: a survey, *Phys. D* **112**, pp. 1–39 (1998).

[74] J. LaSalle and S. Lefschetz, *Stability by Liapunov's Direct Method, with Applications, Mathematics in Science and Engineering*, Vol. 4. Academic Press, New York-London (1961).

[75] V. Lebedev, *An Introduction to Functional Analysis in Computational Mathematics*. Birkhäuser Boston, Inc., Boston, MA (1997).

[76] H. Leiva, Dynamical spectrum for scalar parabolic equations, *Appl. Anal.* **76**, pp. 9–28 (2000).

[77] F. Lettenmeyer, Über das asymptotische Verhalten der Lösungen von Differentialgleichungen und Differentialgleichungssystemen. Verlag d. Bayr. Akad. d. Wiss. (1929).

[78] B. Levitan and V. Zhikov, *Almost Periodic Functions and Differential Equations*. Cambridge University Press, Cambridge-New York (1982).

[79] X.-B. Lin, Exponential dichotomies and homoclinic orbits in functional-differential equations, *J. Differential Equations* **63**, pp. 227–254 (1986).

[80] Z. Liu and R. Yuan, Stability and bifurcation in a harmonic oscillator with delays, *Chaos Solitons Fractals* **23**, pp. 551–562 (2005).

[81] M. Lizana, Exponential dichotomy singularly perturbed linear functional differential equations with small delays, *Appl. Anal.* **47**, pp. 213–225 (1992).

[82] A. Lyapunov, *The General Problem of the Stability of Motion*. Taylor & Francis, Ltd., London (1992).

[83] L. Magalhães, *The spectrum of invariant sets for dissipative semiflows, Dynamics of Infinite-Dimensional Systems (Lisbon, 1986), NATO Adv. Sci. Inst. Ser. F Comput. Systems Sci.*, Vol. 37. Springer, Berlin, pp. 161–168 (1987).

[84] A. Maĭzel', On stability of solutions of systems of differential equations, *Ural. Politehn. Inst. Trudy* **51**, pp. 20–50 (1954).

[85] R. Mañé, *Lyapounov exponents and stable manifolds for compact transformations, Geometric dynamics (Rio de Janeiro, 1981), edited by J. Palis, Lecture Notes in Mathematics*, Vol. 1007. Springer, Berlin, pp. 522–577 (1983).

[86] J. Massera and J. Schäffer, Linear differential equations and functional analysis. I, *Ann. of Math. (2)* **67**, pp. 517–573 (1958).

[87] J. Massera and J. Schäffer, *Linear Differential Equations and Function Spaces, Pure and Applied Mathematics*, Vol. 21. Academic Press, New York-London (1966).

[88] K. Matsui, H. Matsunaga and S. Murakami, Perron type theorem for functional differential equations with infinite delay in a Banach space, *Nonlinear Anal.* **69**, pp. 3821–3837 (2008).

[89] P. McSwiggen, A geometric characterization of smooth linearizability, *Michigan Math. J.* **43**, pp. 321–335 (1996).

[90] A. Mielke, A reduction principle for nonautonomous systems in infinite-dimensional spaces, *J. Differential Equations* **65**, pp. 68–88 (1986).

[91] N. Minh and N. Huy, Characterizations of dichotomies of evolution equations on the half-line, *J. Math. Anal. Appl.* **261**, pp. 28–44 (2001).

[92] N. Minh, F. Räbiger and R. Schnaubelt, Exponential stability, exponential expansiveness, and exponential dichotomy of evolution equations on the half-line, *Integral Equations Operator Theory* **32**, pp. 332–353 (1998).

[93] Yu. Mitropolsky, A. Samoilenko and V. Kulik, *Dichotomies and Stability in Nonautonomous Linear Systems, Stability and Control: Theory, Methods and Applications*, Vol. 14. Taylor & Francis, London (2003).

[94] J. Moser, On a theorem of Anosov, *J. Differential Equations* **5**, pp. 411–440 (1969).

[95] A. Myshkis, *Lineare Differentialgleichungen mit nacheilenden Argument* (transl. from the Russian, 1951). Deutscher Verlag. Wiss. Berlin (1955).

[96] R. Naulin and M. Pinto, Admissible perturbations of exponential dichotomy roughness, *Nonlinear Anal.* **31**, pp. 559–571 (1998).

[97] G. Orosz and G. Stépán, Hopf bifurcation calculations in delayed systems with translational symmetry, *J. Nonlinear Sci.* **14**, pp. 505–528 (2004).

[98] G. Orosz and G. Stépán, Subcritical Hopf bifurcations in a car-following model with reaction-time delay, *R. Soc. Lond. Proc. Ser. A Math. Phys. Eng. Sci.* **462**, pp. 2643–2670 (2006).

[99] V. Oseledets, A multiplicative ergodic theorem. Liapunov characteristic numbers for dynamical systems, *Trans. Moscow Math. Soc.* **19**, pp. 197–221 (1968).

[100] J. Palis, On the local structure of hyperbolic points in Banach spaces, *An. Acad. Brasil. Ci.* **40**, pp. 263–266 (1968).

[101] K. Palmer, Exponential dichotomies and transversal homoclinic points, *J. Differential Equations* **55**, pp. 225–256 (1984).

[102] K. Palmer, Transversal heteroclinic points and Cherry's example of a non-integrable Hamiltonian system, *J. Differential Equations* **65**, pp. 321–360 (1986).

[103] A. Pazy, *Semigroups of Linear Operators and Applications to Partial Differential Equations*, Applied Mathematical Sciences, Vol. 44. Springer-Verlag, New York (1983).

[104] O. Perron, Über Stabilität und asymptotisches Verhalten der Integrale von Differentialgleichungssystemen, *Math. Z.* **29**, pp. 129–160 (1929).

[105] O. Perron, Die Stabilitätsfrage bei Differentialgleichungen, *Math. Z.* **32**, pp. 703–728 (1930).

[106] Ya. Pesin, Families of invariant manifolds corresponding to nonzero characteristic exponents, *Math. USSR-Izv.* **10**, pp. 1261–1305 (1976).

[107] Ya. Pesin, Characteristic Ljapunov exponents, and smooth ergodic theory, *Russian Math. Surveys* **32**, pp. 55–114 (1977).

[108] Ya. Pesin, Geodesic flows on closed Riemannian manifolds without focal points, *Math. USSR-Izv.* **11**, pp. 1195–1228 (1977).

[109] É. Picard, *La mathématique dans ses rapports avec la physique*, Atti del IV Congresso Intern. dei Matemat. (Rome, 1908), Rome, pp. 183–195 (1909).

[110] M. Pituk, Asymptotic behavior and oscillation of functional differential equations, *J. Math. Anal. Appl.* **322**, pp. 1140–1158 (2006).

[111] M. Pituk, A Perron type theorem for functional differential equations, *J. Math. Anal. Appl.* **316**, pp. 24–41 (2006).

[112] V. Pliss, A reduction principle in the theory of stability of motion (in Russian), *Izv. Akad. Nauk SSSR Ser. Mat.* **28**, pp. 1297–1324 (1964).

[113] V. Pliss and G. Sell, Robustness of exponential dichotomies in infinite-dimensional dynamical systems, *J. Dynam. Differential Equations* **11**, pp. 471–513 (1999).

[114] O. Polossuchin, *Über eine besondere Klasse von differentialen Funktionalgleichungen*, Inaugular Dissertation, Zürich (1910).

[115] L. Popescu, Exponential dichotomy roughness on Banach spaces, *J. Math. Anal. Appl.* **314**, pp. 436–454 (2006).

[116] P. Preda, A. Pogan and C. Preda, (L^p, L^q)-admissibility and exponential dichotomy of evolutionary processes on the half-line, *Integral Equations Operator Theory* **49**, pp. 405–418 (2004).

[117] C. Pugh, On a theorem of P. Hartman, *Amer. J. Math.* **91**, pp. 363–367 (1969).

[118] J. Roberts and R. Quispel, Chaos and time-reversal symmetry. Order and chaos in reversible dynamical systems, *Phys. Rep.* **216**, pp. 63–177 (1992).

[119] H. Rodrigues and J. Ruas-Filho, Evolution equations: dichotomies and the Fredholm alternative for bounded solutions, *J. Differential Equations* **119**, pp. 263–283 (1995).

[120] H. Rodrigues and M. Silveira, Properties of bounded solutions of linear and nonlinear evolution equations: homoclinics of a beam equation, *J. Differential Equations* **70**, pp. 403–440 (1987).

[121] R. Sacker and G. Sell, A spectral theory for linear differential systems, *J. Differential Equations* **27**, pp. 320–358 (1978).

[122] R. Sacker and G. Sell, Dichotomies for linear evolutionary equations in Banach spaces, *J. Differential Equations* **113**, pp. 17–67 (1994).

[123] E. Schmidt, Über eine Klasse linearer funktionaler Differentialgleichungen, *Math. Ann.* **70**, pp. 499–524 (1911).

[124] G. Sell, Smooth linearization near a fixed point, *Amer. J. Math.* **107**, pp. 1035–1091 (1985).

[125] G. Sell and Y. You, *Dynamics of Evolutionary Equations*, Applied Mathematical Sciences, Vol. 143. Springer-Verlag, New York (2002).

[126] M. Sevryuk, *Reversible Systems*, Lecture Notes in Mathematics, Vol. 1211. Springer-Verlag, Berlin (1986).

[127] J. Shin, Existence of solutions and Kamke's theorem for functional-differential equations in Banach spaces, *J. Differential Equations* **81**, pp. 294–312 (1989).

[128] J. Sijbrand, Properties of center manifolds, *Trans. Amer. Math. Soc.* **289**, pp. 431–69 (1985).

[129] N. Sternberg, *A Hartman-Grobman theorem for maps, Ordinary and Delay Differential Equations (Edinburg, TX, 1991)*, Pitman Res. Notes Math. Ser., Vol. 272. Longman Sci. Tech., Harlow, pp. 223–227 (1992).

[130] N. Sternberg, A Hartman-Grobman theorem for a class of retarded fuctional differential equations, *J. Math. Anal. Appl.* **176**, pp. 156–165 (1993).

[131] S. Sternberg, Local contractions and a theorem of Poincaré, *Amer. J. Math.* **79**, pp. 809–824 (1957).

[132] S. Sternberg, On the structure of local homeomorphisms of euclidean *n*-space. II., *Amer. J. Math.* **80**, pp. 623–631 (1958).

[133] D. Tucker, A note on the Riesz representation theorem, *Proc. Amer. Math. Soc.* **14**, pp. 354–358 (1963).

[134] N. Van Minh, F. Räbiger and R. Schnaubelt, Exponential stability, exponential expansiveness, and exponential dichotomy of evolution equations on the half-line, *Integral Equations Operator Theory* **32**, pp. 332–353 (1998).

[135] A. Vanderbauwhede, *Centre manifolds, normal forms and elementary bifurcations, Dynamics Reported Ser. Dynam. Systems Appl.*, Vol. 2. Wiley, Chichester, pp. 89–169 (1989).

[136] A. Vanderbauwhede and G. Iooss, *Center manifold theory in infinite dimensions, Dynamics Reported Expositions Dynam. Systems (N.S.)*, Vol. 1. Springer, Berlin, pp. 125–163 (1992).

[137] A. Vanderbauwhede and S. van Gils, Center manifolds and contractions on a scale of Banach spaces, *J. Funct. Anal.* **72**, pp. 209–224 (1987).

[138] V. Volterra, *L'applicazione del calcolo ai fenomeni di eredità*, Revue du Mois, 1912.

[139] M. Wojtkowski, Invariant families of cones and Lyapunov exponents, *Ergodic Theory Dynam. Systems* **5**, pp. 145–161 (1985).

[140] J. Wu, *Theory and Applications of Partial Functional Differential Equations, Applied Mathematical Sciences*, Vol. 119. Springer-Verlag, New York (1996).

[141] W. Zeng, Transversality of homoclinic orbits and exponential dichotomies for parabolic equations, *J. Math. Anal. Appl.* **216**, pp. 466–480 (1997).

[142] W. Zhang, The Fredholm alternative and exponential dichotomies for parabolic equations, *J. Math. Anal. Appl.* **191**, pp. 180–201 (1985).

Index

Printed in the United States
by Baker & Taylor Publisher Services